D0674247

COAL INFORMATION SOURCES DATA BASES

COAL INFORMATION SOURCES AND DATA BASES

by

Carolyn C. Bloch

NOYES DATA CORPORATION

Park Ridge, New Jersey, U.S.A.

1980

Library of Congress Catalog Card Number: 80-22344
ISBN: 0-8155-0830-1
Printed in the United States

Published in the United States of America by
Noyes Data Corporation
Noyes Building, Park Ridge, New Jersey 07656

Library of Congress Cataloging in Publication Data

Bloch, Carolyn.
Coal information sources and data bases.

Includes index.
1. Coal--Information services--Directories.
I. Title.
TN800.B57 553.2'4'07 80-22344
ISBN 0-8155-0830-1

Foreword

This directory comprises a comprehensive guide to federal, state, and international agencies, departments, offices and cooperating information sources, such as libraries and universities, that deal in some capacity with coal data, whether it be legal and regulatory data, assistance programs, research and technology, etc.

Determining the proper organization as a source of energy information can be as perplexing as making use of the information once it is obtained. All of these organizations amass and store vast amounts of data. Successful retrieval of this information can serve as an economical bibliographic tool for scientists, managers and engineers.

This time-saving directory of information sources and data bases supplies the address and a capsule description of each source, delineating its field of emphasis, services offered and, many times, availability of publications aside from the departmental reports which are usually provided free of charge.

The reader will want to make use of both the expanded Table of Contents which precedes the text and serves as a guide to subject matter, and the Index which provides a full listing of all the coal information and data sources.

Table of Contents

Introduction

The United States has a tremendous opportunity to increase the use and production of coal to meet short- and long-term energy goals. The complexity and numbers of policies and regulations presently influence the use and production of coal and will affect the industry for many years.

There are presently over thirty Federal government agencies and departments that have jurisdiction over the production, leasing, mining, research activities, technology, transportation, and tax policies that affect the use and production of coal. Environmental concerns, social impacts and occupational hazards also present great problems for the coal industry. These same agencies however also provide timely valuable information that will help the coal users and producers deal with all of these concerns. Legislative committees, state agencies, and the private sector also present timely data that can be of tremendous value in solving the problems presently inherent in the industry.

The purpose of this directory is to pinpoint all of the exact coal data sources and data bases in order to produce time-saving material that will be of great assistance to scientists, engineers, geologists, managers and other individuals directly involved with the usage of coal data and information. Rapid retrieval of this information can only be accomplished with a complete understanding of coal data location.

The information for the directory was compiled from a large number of sources. Computerized data, Federal directories, and personal contacts with individuals were of tremendous assistance within the public sector. State agencies, associations, universities, international agencies and banks also proved most helpful in providing information on services offered and on publications.

Since the energy field is so rapidly changing, the informational sources will also continue to change and increase. However, this directory covers all the principal coal information sources and these sources will help individuals, companies and organizations locate new coal data as it appears within the public and private sectors.

General Sources

DEPARTMENT OF COMMERCE

NATIONAL TECHNICAL INFORMATION SERVICE
Department of Commerce
5285 Port Royal Road
Springfield, VA 22161
Telephone: (703) 487-4650 New orders for documents and reports
(703) 487-4630 Subscriptions
(703) 487-4763 Computer Products
(703) 487-4642 NTISearches

Description: NTIS is the central source for the public sale of U.S. and foreign government-sponsored research, development and engineering reports, and other analyses prepared by national and local government agencies, their contractors or grantees or by Special Technology Groups.

Products that provide individuals assistance in order to keep up to date on NTIS reports are as follows:

Government Reports Announcement and Index — This is a summary of U.S. government research published in one biweekly volume with an index. Indexed by subject, personal and corporate author, government contract and report/accession numbers. The items are mostly subscription publications and services.

NTIS Bibliographic Data File on Magnetic Tape — Contains research summaries and other data and analysis. Most items have full bibliographic citations and may be used to create a wide variety of information products.

Subscribers may search any of the file elements including research reports, title, personal or corporate author, accession number or contract number and by subject using keywords, descriptors or subject codes.

Information on energy keyword descriptors and information sources is available from "Energy Microthesaurus–A Hierarchical List of Indexing Terms Used by NTIS." This publication will assist those who search NTIS Bibliographic Data File. Also available is "Environmental Microthesaurus—A Hierarchical List of Indexing Terms Used by NTIS."

Published Searches – These are completed, printed and published in response to previous demand or to NTIS anticipation of wide interest in a particular subject area.

Each of the published searches consists of as many as 100 different research summaries of advanced technology. Each report includes abstracts, title of full report, its author, corporate or sponsoring source, pages, price and instructions for ordering the full text.

Users of published searches may multiply their search benefits and applications where NTISearch subjects are offered with identical subject searches from the data bases of Engineering Index and American Petroleum Institute.

Custom Searches (NTISearch) – These are on-line searches of the entire collection in collaboration with an NTIS specialist. NTISearch provides access to some 680,000 reports averaging 250 words of description.

SRIM (Selected Research in Microfiche) – Individuals may choose from 500 subject categories and search for complete research reports in microfiche in only the subject areas selected. SRIM has a quarterly index in paper or microfiche.

Subscription services that provide summaries in specialized fields:

Tech Notes – This is a biweekly package containing page summaries of new applications for technology as developed by Federal agencies and their contractors.

Tech Notes includes the following: NASA Tech Briefs, Army DARCOM Tech Notes, AF Abstracts of New Technology, AF Manufacturing Technology, Navy Manufacturing Technology, Navy Technology Transfer Sheets, Interior—Bureau of Mines Technology News, Agriculture—Forest Service Equipment Tips, Commerce—NBS Staff Reports, DOE-Union Carbide Bulletin, NSF/RANN-UTS Briefs.

The subject categories include computers, electrotechnology, energy, engineering, life sciences, physical sciences, machinery, materials, manufacturing, ordnance and testing and instrumentation.

Newsletters — The newsletters contain research summaries that are within 3 weeks of their receipt by NTIS. Covers 33 areas of government research. Also provides cumulative monthly indexes which give the title of report with the page number, noting where it appeared in the abstract newsletter. The final issue of the year is a special subject index containing up to 10 cross references for each research summary index.

Coal-related newsletters include Building Industry Technology, Business and Economics, Chemistry, Energy, Combustion, Engines and Propellants, Environmental Pollution and Control, Natural Resources and Earth Sciences.

Periodicals — NTIS periodicals include the following:

Energy — A continuing bibliography with Indexes, published quarterly. Coverage includes regional, national and international energy systems, R&D on fuels and other sources of energy, energy conversion, transport, transmission, distribution and storage.

Energy Conservation Update — Provides abstracts and an index to the latest studies of energy conservation. Covers a broad spectrum from transportation and industrial energy savings to the latest research dealing with techniques for residential and commercial heating, lighting and hot water conservation. Available monthly.

Fossil Energy Update — Abstracts information in subject areas such as coal, petroleum, natural gas, oil shale, hydrogen production, hydrocarbon and alcohol fuels, electric power engineering and magnetohydrodynamic generators. Available monthly.

Directory of Computer Software Application: Energy — Presents abstracts of energy-related reports and computer programs developed from federally sponsored research. The abstracts describe in detail the computer software available in fossil, solar, nuclear, geothermal, ocean thermal and other energy areas.

Energy SOS — Contains full text of every printed report published by NTIS in the field of energy. Subjects include energy sources, energy use, supply and demand, power and heat generation, energy conversion and storage, energy transmission, fuel conversion processes, policies, regulations and studies, engines and fuels.

Reports on Energy Information Reported to Congress — The Office of Data and Analysis in the Department of Energy reports to Congress the status of international energy exploration, consumption, storage, production and transportation. Issued quarterly.

Monthly Energy Review — Principal communications medium for EIA. The major parts are an overview summarizing events of the previous month and sections on crude oil and refined petroleum products, natural gas, coal, nuclear power, electric utilities and consumption, and resource development.

EPA Publications Bibliography Quarterly Abstract Bulletin — EPA items in the NTIS collection are published quarterly in a continuing series.

Selected Water Resources Abstracts — Covers water pollution, waterlay, groundwater, lakes, water yield, watershed protection, waste treatment, water demand, hydraulics and soil mechanics.

BUREAU OF CENSUS
Department of Commerce
Washington, DC 20233
Telephone: (301) 568-1200

Description: The Bureau of Census collects, tabulates and publishes a wide variety of statistical data. The results of the census of the mineral industry are published in the "1977 Census of Mineral Industries." The results of the census are presented in a series of reports on industries, geographic areas, subjects and types of organizations. General statistics on the number of establishments, employment, payroll, workhours, cost of materials, value of shipments, and capital expenditures. Statistics are given on quantity and value of materials consumed and products shipped.

The Bureau of Census maintains data files that can be processed to provide subject cross classifications and area tabulations. The Customer Services Branch of the Data User Services Division handles requests for data products. The Branch provides consultation and analytical services to users needing assistance.

DEPARTMENT OF ENERGY

ENERGY INFORMATION ADMINISTRATION
Department of Energy
1000 Independence Avenue SW
Washington, DC 20585
Telephone: (202) 252-8800

Description: Collects, processes and publishes data on energy reserves, production, demand, consumption, the financial status of energy producing companies and other areas. Provides analyses of data to assist individuals in both the private and public sectors in understanding energy trends.

EIA publications directory is available through the clearinghouse or through GPO. Publications and reports are available in the following categories: general, petroleum, coal minerals, natural gas, electricity, and other energy sources. Special reports provide information to meet short-term data requirements. Publications include program summaries, publications on applied analysis, energy data, energy information sources, and program development. Available on microfiche and will be available on computer tape.

Offices with coal data within EIA:

> Coal and Electric Power Statistics Division within the Office of the Assistant Administrator for Energy Data Operations collects, compiles and evaluates data on coal and the electric power industries. The Office issues reports and publications pertaining to coal reserves, production and distribution, coke and coal chemicals, electric retail rates, power disturbances, electricity production and sales, fuel consumption and stocks at generating facilities, and import and export figures.
>
> Coal and Electric Power Analysis Division within the Office of the Assistant Administrator for Applied Analysis provides information on coal mining equipment, coal economic analysis, supply and demand. Studies and produces reports on export figures for coal.
>
> National Energy Information Center within the Office of Energy Information Services provides comprehensive sources of information about energy data. Access to catalogs of energy information products, systems, data bases, and models incorporated in NEIC. Provides compilations of energy data. Answers inquiries and provides expertise on energy sources and applications. Provides on-line searches and retrieval access to both government and commercial data bases.

TECHNICAL INFORMATION CENTER
P.O, Box 62
Oak Ridge, TN 37830
Telephone: (615) 576-1188

Description: Maintains on a current basis a comprehensive bibliographic data base on energy. The data bases can be accessed by DOE RECON.

The Energy Data Bases within TIC contain the following subject categories:

Fossil and Synthetic Fuels — Current citations included in the data base are coal, petroleum, natural gas, oil shales, tar sands, hydrogen production, hydrocarbon fuels, alcohol fuels, electric power engineering, MHD, generators and combustion, chemistry and systems.

Specifically the coal data base contains information on processing (carbonization, desulfurization, hydrogenation, gasification, liquefaction, pyrolysis and solvent extraction), by-products, properties, waste management, environmental aspects, reserves and exploration, mining, transport and handling, preparation, combustion, marketing and economics, health and safety, and regulations. Bibliographies published include *Coal Processing 1975, Coal Processing Production and Properties 1976.*

Energy Policy and Management Data Base — Information includes legislation, regulation and energy management on energy, systems, regional analyses, economics, sociology and politics, environment, health and safety, resource assessment and conservation, R&D programs, supply, demand, forecasting, consumption and utilization.

Environment Sciences Data Base — Contains information on policies for mitigating or eliminating adverse effects. All energy technologies are covered including conservation, coal oil shale, petroleum, natural gas, solar, geothermal, fusion and fission. Special information is included on the identification and analysis of hazardous products from energy technologies such as coal conversion.

Engineering Data Base — Includes information on engineering studies related to facilities and equipment, materials, waste processing, combustion systems, and underground engineering.

Research in Progress File — Designed to provide a record of DOE research and projects funded by other government agencies—federal and state, private, and industrial research organizations. Each project contains research sponsoring organizations, a summary description of the project, subject indexing terms, and categorization. Information is machine searchable.

DEPARTMENT OF TRANSPORTATION

TRISNET
Research and Special Program Administration
Department of Transportation
400 7th Street SW
Washington, DC 20590
Telephone: (202) 426-0975

Description: Trisnet links the transportation information activities into a system of libraries, data bases and retrieval services. Data include abstracts of transportation literature, photocopies of reports, and reports on planned and ongoing research, references to numerical data bases, directories to transportation information centers, and statistical data.

Trisnet provides data on rail transport, economics, fuel, energy and safety. Tris on-line is a computerized information retrieval system of selected references to technical literature and ongoing research résumés available from remote computer terminals.

Transportation research information resources that provide data pertinent to coal through Trisnet include:

> *Railroad Research Information Service* — Sponsored by the Federal Railroad Administration. Subject areas include rail transport, public transportation, safety, engineering, operations, and economics. Produces *Railroad Research Bulletin.* Printouts of batch mode and on-line retrievals on specific topics are available from RRIS and Tris on-line data bases.
>
> *Transportation Research Board Library* — Subjects include all modes of transportation. *Transportation Research Abstracts* is available monthly.
>
> *Transportation Systems Center* — Located in Cambridge, MA; provides information on transportation, socio-eco engineering, science and technology, energy and operations. Produces *Technical Information Center Bulletin.*

DEPARTMENT OF DEFENSE

DEFENSE TECHNICAL INFORMATION CENTER

Reference Section
Cameron Station
Alexandria, VA 22314
Telephone: (202) 274-7633

Description: The Defense Technical Information Center is the clearinghouse for DOD collection of R&D in all fields of science and technology. Contains information as related to research, development, test and evaluation. Contains information on what research is planned, performed and the completed results. Department of Defense and associated contract researchers deposit information into data banks collected by the Defense Technical Information Center.

Data Banks:

R&DPP — Repository of program planning documentation at the project and task level.

R&T Work Unit Information System — Collection of technically oriented summaries describing research and technical projects currently in progress at the work unit level. Information is available concerning what, where, when, how, at what costs, by whom and under whose sponsorship research is performed.

Technical Report Data Bank — Collection of formally documented science and technical results of Department of Defense sponsored research, development, test and evaluation. Products, programs and services are available from the technical report collection in bibliographies, and taped distribution.

Independent R&D Data Bank — Data bank of information describing the technical progress being performed by DOD contractors as part of their independent R&D programs.

Specific data banks with information pertinent to the coal industry:

Chemistry — Chemical engineering and physical chemistry.

Earth Sciences — Geochemistry, geodesy, geography, geology and mineralogy, mining engineering.

Energy Conversion — Conversion techniques, power sources, energy storage.

The other data banks are aeronautics, agriculture, astronomy and astrophysics, atmospheric sciences, behavioral and social sciences, biological and medical sciences, electronics and electrical engineering, materials, mechanical, industrial, civil and marine engineering, mathematical sciences, methods and equipment, military sciences, missile technology, navigation, communications, detection and countermeasures, nuclear science and technology, ordnance, physics, propulsion and fuels, and space technology.

By-products of the Data Banks:

Announcement publications
Automatic Document Distribution Program
Automatic Magnetic Tape Distribution Program
Bibliography Program
Defense R&D of the 1960s
Defense R&D of the 1970s
Selective Dissemination of Information
Technical Vocabulary
Recurring Management Information Systems Reports

Other Services Offered:

Defense Technical Information Center Referral Service – Data bank of government-sponsored activities specializing in scientific and technical information not available at the center.

Central Registry – Central file of users with authorized access to DOD science and technology information.

DOD RDT&E On-Line System – Network of remote terminal stations linked to the Center's central computer for instant visual display of data from four major collections.

Announcements of the Center's programs, products and services:

Technical Abstract Bulletin
Technical Abstract Bulletin Index
Annual Indexes
TAB Quarterly Indexes
Bibliography of Bibliographies

Eligibility Requirements:

Individuals that have R&D activities within the Federal government and their associated contractors, subcontractors and grantees with current contracts may be eligible for service by military service authorization under DOD potential contractor.

Registration:

Registration for the Center's services assists the user in obtaining services offered by Defense-sponsored Information Analysis Centers and major technical libraries.

Registration Procedures:

Requester will be sent manual entitled "Registration for Science and Technical Information Services of Defense." This manual explains the procedures. Requester will receive a DOD user code number and be placed on a central registry. Requester will receive a duplicate copy of DD form 1540 with appropriate materials.

Ordering Publications:

Two options are open to individuals: (1) Order directly through NTIS with check, or money order. Some charge cards are accepted with NTIS. (2) Order directly from the Center or NTIS after establishment of a NTIS Deposit Account. Defense Technical Information Center users are encourgaged to open a NTIS Deposit Account since faster service can be provided.

NASA

NASA SCIENTIFIC AND TECHNICAL INFORMATION SYSTEM

400 Maryland Avenue, SW
Washington, DC 20546
Telephone: (202) 755-3548

Description: The system supports the NASA R&D efforts and assists in the dissemination of NASA program-generated information to the public. The system covers all the sciences and technologies related to aeronautics, space-earth resource surveys, solar and wind energy and other areas of NASA R&D.

The Scientific and Technical Aerospace Reports (STAR) contain information on science and technology as related to space. STAR announces reports, doctoral theses, NASA patents and includes current research projects. Publications from internal sources issued through STAR include Contractors' Reports, Technical Memoranda, and Technical Notes and Reports.

NASA/SCAN covers announcements of selected STAR and IAA citations in over 200 subject topics available to NASA and contractors.

SMITHSONIAN SCIENCE INFORMATION EXCHANGE

SMITHSONIAN SCIENCE INFORMATION EXCHANGE

1730 M Street NW
Washington, DC 20036
Telephone: (202) 634-3933

Description: SSIE presently is a nonprofit corporation that assists in the planning and performance of research activities by providing up-to-date information about research in progress. The Exchange collects, indexes, stores and disseminates data about basic and applied work in all areas of the life and physical sciences.

SSIE Search Services:

> *Custom Searches* – Staff scientists search the active files for Notices of Research Projects on specific subjects. Searches for NRP's from particular performing organizations or departments, specific geographic areas or any combination of similar requirements can also be made. The fee for this service is $50.00 for the search and the first 50 project notices sent.
>
> In conjunction with NTIS, SSIE offers combined custom searches of the SSIE file and of the NTIS file of abstracts from technical reports on federally sponsored research. The fee for combined custom searches is $30.00 for the search service and the first 150 documents sent.

Research Information Packages — SSIE scientists conduct searches of the active file for NRP's on subjects of high current interest. These RIP's are announced in the SSIE Science Newsletter. Packages are updated at least every 90 days. Prices vary with 1 to 25 NRP's $25.00; 25 to 100 NRP's $35.00; 101 to 200 NRP's $45.00; 201 to 300 NRP's $55.00.

Coal-Related NRP's:

Strip Mining	LM 25	$45.00
Coal Mine Safety	LM 20	$75.00
Coal Mining Equipment and Techniques	LM 19	$65.00
Western Coal	LB 12	$55.00
Coal Mine Wastes	LA 97	$45.00
Coal Mining Effects on Water Quality	LA 95	$45.00
Coal-Fired MHD	IB 11	$35.00
Trace Elements in Coal	FB 19	$45.00
Catalyst for Coal Conversion	FB 24	$55.00
Coal Desulfurization	FN 11	$45.00
Coal Gasification	FN 16	$75.00
Coal Liquefaction	FN 17	$55.00
Fluidized Bed Combustion	FN 24	$35.00
Energy Sources, Supplies and Demand (Coal Included)	CB 02	$55.00
Coal Derivatives for Electric Power Generation	FB 04	$35.00

Selective Dissemination of Information — SSIE scientists establish user interest profiles for each subscriber. Periodic searches of the active file are conducted according to profile requirements to identify all new or newly updated project notices added to the data base since a previous search was made.

SDI Services:

Standard SDI Service — Subscribers receive 12 monthly search updates compiled for the single fee of $180.00.

Custom SDI Service — Subscribers are provided with quarterly updates, each of which is received by a staff scientist to assure maximum relevance of update contents to search requirements.

SSIE Science Newsletter — Each issue contains newest research information package titles plus articles of interest to the scientific community.

On-Line Search Service — The SSIE data base is available online for users with access to a computer terminal.

ELECTRIC POWER RESEARCH INSTITUTE

ELECTRIC POWER RESEARCH INSTITUTE
3412 Hillview Avenue
P.O. Box 10412
Palo Alto, CA 94303
Telephone: (415) 855-2000

Description: EPRI was founded in 1972 by the major sectors of the nation's utility industry to develop and administer a coordinated national electric power research and development program. Through selection funding and management of research projects conducted by contracting organizations, EPRI promotes the development of new and improved technologies to help meet present and future electric energy needs.

Primary areas of research are fossil fuel systems, advanced technology systems, nuclear power, electrical systems, environmental assessment and energy analysis. Reports are published in 38 program areas with six divisions as follows:

Advanced Power Systems Division,
Coal Combustion Systems Division,
Electrical Systems Division,
Energy Analysis and Environment Division,
Energy Management and Utilization Division, and
Nuclear Power Division.

Technical reports describing R&D work sponsored by EPRI are available for purchase. EPRI publishes "EPRI Guide." This is a directory of technical reports, audiovisual materials, computer programs and data bases, licensable inventions, R&D information systems and other information services.

Libraries

DEPARTMENT OF ENERGY

DEPARTMENT OF ENERGY LIBRARY
1000 Independence Avenue SW
Washington, DC 20585
Telephone: (202) 252-9534

Description: Contains monograph collection of approximately 40,000 titles. The microfiche collection has close to one-half million documents and the journal collection contains approximately 2,300 titles. Provides on-line searching on DOE RECON, Lockheed, SDC, JURIS, MEDLINE, Dow Jones, NASA RECON, BRS, OCLC and the New York Times Information Systems.

A bibliography issued periodically contains listings of DOE headquarters publications. Three types of publications are available. These are (1) publications that deal with programs and policy; (2) reports prepared by contractors and published by DOE to describe R&D work performed for DOE; and (3) environmental development plans and impact statements.

Sources for DOE publications are as follows:

Economic Regulatory Administration
Room B-110
2000 M Street NW
Washington, DC 20461

FERC
Room 1000
825 North Capitol Street NE
Washington, DC 20426

GPO
Superintendent of Documents
Washington, DC 20402

National Energy Information Center
Room 850
1726 M Street NW
Washington, DC 20461

NTIS
U.S. Department of Commerce
Springfield, VA 22161

DOE Technical Information Center
P.O. Box 62
Oak Ridge, TN 37830

DEPARTMENT OF ENERGY LABORATORY LIBRARIES

Publications available at the DOE Headquarters library are also available at some regional locations. Copies of DOE Environmental Assessments and of Draft Environmental Impact Statements, along with comment letters and other materials related to the EIS may be seen in the DOE public reading room and at selected public document rooms throughout the country.

Ames Laboratory Library
Iowa State University
Department of Energy
Ames, IA 50011
Telephone: (515) 294-1856

Description: The Document library deals primarily in R&D reports, research notebooks and papers published by the laboratory staff. Subject areas include coal R&D.

Argonne National Laboratory Library
Technical Information Services Department
Department of Energy
9700 S Cass Avenue
Argonne, IL 60439
Telephone: (312) 972-4221

Description: Main subject areas are nuclear science and engineering, chemistry, biological sciences, energy and environmental sciences, materials science, mathematics, physics, coal and coal usage. The National Energy Software Center is located at the laboratory.

Brookhaven National Laboratory Library
Department of Energy
Upton, Long Island, NY 11973
Telephone: (516) 345-2123

Description: Information is available on energy resources, management, and environmental studies. The Research Library contains information on chemistry, energy, engineering, and environmental sciences.

INEL Technical Library
EG&G Idaho Inc.
Department of Energy
P.O. Box 1625
Idaho Falls, ID 83401
Telephone: (208) 526-1194

Description: Library holdings include nuclear science and such fields as chemistry, engineering, instrumentation, computer science and metallurgy. In recent years the collection has broadened to include all fields of energy R&D.

Lawrence Livermore Laboratory Library
Department of Energy
University of California
P.O. Box 808
Livermore, CA 94550
Telephone: (415) 422-9310

Description: The laboratory conducts research programs in coal gasification. The technical information department has data on energy, environment, physics, chemistry and general reference.

Oak Ridge National Laboratory Library
P.O. Box X
Oak Ridge, TN 37830
Telephone: (615) 574-7178

Description: The library is part of the Information Division of the laboratory. Subject areas include biology, chemistry, physics, nuclear technology, metallurgy, environmental science and energy.

The library provides translation services, SDI services and computerized retrieval services including RECON, Lockheed DIALOG, SDC ORBIT, and N.Y. Times Information Bank.

The component centers of the Information Division of the laboratory provide major services such as bibliographic references, organization of collected materials and assessment of information. Coverage includes information on air quality, chlorination effects, coal conversion effects, coal processing technology, ecological research, energy analysis, energy conservation, energy research and development, environmental exposure hazards, environmental monitoring, environmental sciences research, environmental standards and criteria, wastewater treatment and water quality.

These centers are Ecological Sciences Information Center, Energy and Environmental Resource Center, Energy Research and Development Inventory, Environmental Mutagen Information Center, Environmental Teratology Information Center, Health and Environmental Studies Program, Toxicology Data Bank, and the Toxicology Information Response Center.

There are also information and data centers organized within the research divisions of the laboratory. These centers focus mainly on highly specialized areas of the physical sciences and engineering.

A computerized system (ORLOOK) is located at Union Carbide's Nuclear Division at Oak Ridge National Laboratory. The data bank contains engineering information pertaining to coal systems and technology.

Sandia Laboratories Library 3140, Albuquerque
Department of Energy
Albuquerque, NM 87185
Telephone: (505) 264-8765

Description: Involved in research and development pertaining to fossil fuels. The library contains information on physics, energy sources and environmental engineering. Machine readable data bases are used in house for current awareness service. Two terminals are available for DOE/RECON and Defense Technical Information Center's DRDT&E retrieval systems. Dial up access to NASA's RECON, the New York Times Information Bank, SDC's ORBIT, and Lockheed's DIALOG.

Sandia Laboratories Library, Livermore
Department of Energy
Organization 8266
Livermore, CA 94550
Telephone: (415) 455-2271

Description: Subjects include nuclear ordnance, materials, physics, mathematics, environmental engineering, energy conversion and utilization.

DEPARTMENT OF ENERGY TECHNOLOGY CENTERS LIBRARIES

Grand Forks Energy Technology Center
Department of Energy
Box 8213
University Station
Grand Forks, ND 58202
Telephone: (701) 795-8132

Description: Areas of interest include engineering, physical sciences, mathematics and chemistry as related to the combustion, gasification, liquefaction and emissions control of low rank Western coal and peat. Information is available on mining preparation and storage of lignite.

Publications include Quarterly Technical Progress Report, journal articles and reprints. Included in the holdings are a collection of Bureau of Mines Bulletins, Reports of Investigations, Information Circulars, Minerals Yearbooks, and Mineral Industry Surveys.

Answers inquiries, provides information on research in progress and makes referrals.

Laramie Energy Technology Center
Department of Energy
P.O. Box 3395
University Station
Laramie, WY 82071
Telephone: (307) 721-4201

Description: Information is available on shale, shale oil, coal gasification (underground) tar sands, and asphalt.

Publications include technical reports, journal articles, conference papers and bibliographies.

Answers inquiries, provides reference services, referrals, permits on-site use of collection. Services are free and available to anyone.

Morgantown Energy Technology Center
Department of Energy
P.O. Box 880
Morgantown, WV 26505
Telephone: (304) 599-7183

Description: Areas of interest include development of technology equipment and materials for processes for producing gas, oil and electricity from coal. Information is available on the extraction of oil and gas, coal feeding, coal gasification, coal combustion, gas cleanup, fluid bed combustions, earth fracture systems, underground gasification, and secondary and tertiary oil and gas recovery.

Publishes technical reports, journal articles and reprints. A publications list is available.

Answers inquiries, provides information on R&D in progress and makes referrals.

Holdings include literature collection, crude oil and oil field brine analyses, cores and loss from selected Eastern oil field, computerized data base of geology and petroleum engineering associated with Eastern shales.

Pittsburgh Energy Technology Center
Department of Energy
4800 Forbes Avenue
Pittsburgh, PA 15213
Telephone: (412) 675-6000

Description: PETC is developing new processes for producing clean energy from coal. Areas of interest include coal gasification processes, coal liquefaction processes, magnetohydrodynamic power generation, coal preparation, physical and chemical desulfurization, coal combustion, coal chemicals, coal analysis, pollutant removal, coal-oil mixture combustion, conservation and environmental studies of coal conversion processes.

Plans, conducts, and directs R&D programs for converting coal in an environmentally acceptable manner to clean burning liquid fuels as well as conducting an extensive program on various aspects of coal conversion.

Publications include technical reports, journal articles, conference papers.

A microfiche collection has been added covering nonnuclear information in the energy field.

DEPARTMENT OF ENERGY CONTRACTOR ORGANIZATION LIBRARIES

Battelle N.W. Library
Pacific N.W. Laboratory
P.O. Box 999
Richland, WA 99352
Telephone: (509) 942-3443

Description: Subjects offered are nuclear science and engineering, metallurgy, materials, radiological sciences, environmental sciences, chemistry, physics and mathematics. Technical files hold over 300,000 reports, hard copy and microfiche.

Bituminous Coal Research Library
350 Hochberg Road
Monroeville, PA 15146
Telephone: (412) 327-1600

Description: Serves the needs of BCR. Subject areas include coal mining, coal as a material, mine water, pollution abatement, land reclamation, refuse utilization and disposal and coal using processes such as combustion, gasification and liquefaction.

Chevron Research Company Library
576 Standard Avenue
P.O. Box 1627
Richmond, CA 94802
Telephone: (415) 237-4411

Description: Subject areas include chemistry, energy engineering, environmental sciences, mathematics and physics.

Energy Search Center Library
Texas Industrial Commission
P.O. Box 12728, Capitol Station
Austin, TX 78711
Telephone: (512) 472-5059

Description: Subject areas include energy, statistics, industrial energy, conservation, energy resources and industrial and mechanical engineering. The center offers on-line bibliographic literature searches and has access to DOE/RECON, Lockheed DIALOG, SDC ORBIT, and the N.Y. Times Information Bank Systems.

Fluidyne Engineering Corporation Library
5900 Olson Memorial Highway
Minneapolis, MN 55422
Telephone: (612) 725-4242

Description: Subject areas include aerospace, engineering conservation, fluidized bed and MHD.

Gilbert Associates Library
P.O. Box 1498
Reading, PA 19603
Telephone: (215) 775-2600

Description: Collection includes DOE reports, Bureau of Mines, EPA and other energy related technical reports plus texts related to coal technology and the economic aspects of the energy industry.

Lockheed Palo Alto Research Laboratory Technical Information Center
Lockheed Missiles and Space
3251 Hanover Street
Palo Alto, CA 94304
Telephone: (415) 493-4411

Description: Subjects include aerospace science, defense systems information, electronics and communications sciences, energy conversion, oil recovery, environmental protection, materials sciences, nuclear sciences and ocean energy.

Mason and Hanger-Silas Mason Company Technical Library
Pantex Plant
P.O. Box 30020
Amarillo, TX 79177
Telephone: (806) 335-1581

Description: Subjects are chemical explosives engineering, management, manufacturing, engineering, health and safety engineering, and environmental.

Mitre Corporation Library
1820 Dolley Madison Blvd.
McLean, VA 22102
Telephone: (703) 827-6486

Description: Subjects include systems engineering as applied to energy resources and the environment.

Rockwell International Library
Energy Systems Group
Rocky Flats Plant
P.O. Box 938
Golden, CO 80401
Telephone: (303) 497-4809

Description: Subjects include atomic energy, environment and ecology, energy, metallurgy, chemistry, physics, mathematics, engineering. Limited scope in statistics, management, computer science, and economics. The collection contains 15,000 volumes, 500 journal titles, 30,000 reports, patents, an integrated research file of all formal and informal reports generated on plant site.

DEPARTMENT OF INTERIOR

OFFICE OF LIBRARY AND INFORMATION SERVICES

Office of the Secretary
Department of Interior
18th and C Streets NW
Washington, DC 20240
Telephone: (202) 343-3896

Description: Resource for published materials on natural resources. This includes the scientific, engineering, legal and social aspects of mining and minerals. Provides data for the Department of Interior and researchers from state, regional and local governments.

Computer search services are available to the researchers via a terminal and communications network. Subject areas include chemistry, geology, engineering, physics, patents, pollution and U.S. government R&D reports. Bibliographies and current energy titles are also available. This information is available on periodicals, articles, books and documents.

U.S. GEOLOGICAL SURVEY LIBRARY

National Center
Reston, VA 22092
Telephone: (703) 860-6671

Description: The main library is located in Reston, VA. Branch libraries are located in Denver, CO; Menlo Park, CA; and at Flagstaff, AZ. The libraries contain 900,000 monographs, serials and government publications; 350,000 pamphlets and reprints; 325,000 maps and charts; 12,000 field record notebooks and manuscripts; and 200,000 album prints, transparencies and negatives as well as doctoral dissertations on microfilm and report literature on microfilm.

LIBRARY OF CONGRESS

LIBRARY OF CONGRESS
Washington, DC 20540
Telephone: (202) 287-5000

Description: Coal Information is contained in the Science and Technology Division—Reference Section, Scorpio Information Retrieval System and the National Referral Center for Science and Technology plus other general resources in the library.

Science and Technology Division—Reference – Provides information on library holdings in science and technology. Provides information on technical reports and serial titles. Publishes the series "L.C. Tracer Bulletin." This is a series of literature guides designed to help locate published materials. Coal titles are "74-5 Coal Gasification," "75-8 MHD," and "79-2 Energy Resources in China." "Technology in the Twenty-First Century" is in progress.

The reading room has information on coal gasification, liquefaction, mines, research, coal tar products plus general books on energy.

The energy vertical files are located in the reading room. Coal related files that are available are "Energy and the Environment," "Energy Forecast and Resources," "Energy R&D," and "Fossil Fuels." Answers without charge brief technical inquiries.

Scorpio – The files are available to the Library of Congress and to the users of the library. Terminals are located in the reading rooms. The files that are currently available include:

1. Legislative Information Files – These are a synopsis of the publication entitled "Digest of General Public Bills and Resolutions."

2. Bibliographic Citation File – This is the Congressional Research Services citation file. It is a collection of references to significant current periodical articles, pamphlets, GPO documents, U.N. documents, and interest or lobby group publications.

3. Major Issues File Collection – This is concise objective briefs on key issues of public policy. Each brief contains definition of issues, background and policy analysis statements, references to major legislation, hearings and committee reports, references to professional literature.

4. NRC resource file – This is a collection of more than 10,000 descriptions of information resources.

5. Library of Congress Computerized Catalog – Bibliographic file containing about 600,000 references from the library's MARC data base.

National Referral Center for Science and Technology – This is a free referral service. The referral service uses a subject indexed computerized file called information resources. The center's file which is maintained by professional analysis is used primarily by the center's referral specialists. It is accessible at the library through computer terminals located in the reading room and to federal agencies through RECON computer network operated by DOE.

The NRC Subject Index pertaining to Coal is as follows:

Anthracite Coal
Bituminous Coal
Fossil Fuels
Lignite
Subbituminous Coal
Coal Characterization Data File
Coal and Mineral Deposits
Coal Gas
Coal Gasification
Coal Industry
Mining Leases
Coal Liquefaction
Coal Mines of U.S.
Strip Coal Mining
Coal Mining
Underground Mining
Coal Preparation
Coal Research
Fuel Research
Industrial Research
Coal Reserves
Mineral Reserves
Coal Resources
Nonrenewable Resources
World Natural Resources
Coal Storage
Coal Tar
Coal Technology Data Base
Coal Trade
Coal Transportation
Coal Utilization
Coke
Extractive Metallurgy
Materials Recovery

CONGRESSIONAL RESEARCH SERVICE
Library of Congress
Washington, DC 20540

Description: CRS works exclusively for Congress conducting research, analyzing legislation, and providing information at the request of committee members and staffs, CRS makes such research available in the form of studies, reports, digests and background briefings. Upon request CRS analyzes legislative proposals and issues. CRS assesses the effects of these proposals and their alternatives. Senior specialists and analysts are available for personal consultations.

EXECUTIVE OFFICE OF THE PRESIDENT INFORMATION CENTER

WHITE HOUSE AND EXECUTIVE OFFICE OF THE PRESIDENT INFORMATION CENTER
New Executive Office Building
726 Jackson Place NW
Washington, DC 20503
Telephone: (202) 395-3654

Description: Collections consist of books, journals, newspapers and monographs. Available in hard copy or microfilm. Subject areas include economics and budgetary affairs, management sciences, environmental and natural resources, and science and education policy. Special collections consist of appropriation hearings and legislation, federal budget documents, and legislative histories of reorganization plans. The information center has on-line access to more than 100 computer data bases. These include LEGIS, JURIS, Datasearch, IMS, and Labstat.

SURFACE MINING RESEARCH LIBRARY

SURFACE MINING RESEARCH LIBRARY
Box 5024
Charleston, WV 25311
Telephone: (304) 348-8035

Description: Areas of interest include coal mining, surface mining of coal, cleaning and use of coal and energy competition.

Publications include "Energy and the Environment," "What's the Strip Mining Controversy All About," and "West Virginia Coal Industry and Ways to Help It." Information services include answering inquiries and providing literature search services. Fees are charged for some services.

Holdings include extensive collection of photographs and a collection of slides relating to surface mining, mine reclamation and environmental problems. Contains over 6,000 written documents.

Publications and Information

DEPARTMENT OF INTERIOR

BUREAU OF MINES
Department of Interior
2401 E Street NW
Washington, DC 20240
Telephone: (202) 634-1001

Description: The following types or series of publications are offered:

Bulletins – Bulletins report the results of broad and significant projects or programs of scientific, historical or economic research. They also include investigations including mineral resource studies and compilations. Bulletins are usually prepared after all laboratory and field work has been completed, but sometimes report a major phase of a larger or continuing investigation or research study.

Minerals Yearbook – Annual statistical publication. Summarizes economic and technologic developments in the mineral industry. It presents by mineral commodity the salient statistics on production, trade, consumption and other pertinent data. Mineral statistics are presented for foreign countries along with a review of the role of minerals in the economies of these nations. Three separate volumes are issued each year.

Reports of Investigations – Presents the results of research and investigations conducted at research centers, mines, quarries, smelters, refineries, oil fields and plants. They describe the principal features and results of individual experiments, minor research projects or a significant coordinated phase of a major project or program.

Information Circulars – Covers surveys of mineral resources and related mining and operating activities, guides to marketing and mineral commodities, compilations of historical, statistical and economic data on minerals, summaries of scientific and technical meetings and symposiums, bibliographies, descriptions of new instrumentation and techniques and descriptions of new industrial mining methods and metallurgical processes.

Technical Progress Reports – Presents developments in programs in the form of expanded fact sheets. Gives the technical background and details to supplement press releases.

Mineral Commodity Profiles – Supplements the Minerals Yearbook. Data are presented for each commodity that include background material on industry structure, technology, resources and reserves, economic and production data, and forecasts of future supply/demand relationships and uses.

Mineral Perspectives – Presents latest data on commodities that are of critical importance in a particular foreign country or region. Available on such areas as Canada, Australia, Western Europe, Eastern Europe and the USSR.

State Mineral Profiles – Presents data on mineral resources and production affecting mineral resource development in each state.

Handbooks – Information manuals designed to improve efficiency in the mineral industries.

Mineral Industry Surveys – Contains statistical and economic data on minerals and fuels. Useful for industry to keep informed on trends in production, distribution, inventories and the consumption of minerals and fuels.

Special Mineral Commodity Publications – Provides information that enables domestic producers and consumers of mineral commodities to keep abreast of developments.

Special Publications – Publications that are in response to requests for information on specific subjects.

Computer Tapes and Printouts – Available on mineral data.

Associated Documents – These are cooperative publications, open file reports, outside publications, Bureau of Mines patents and Bureau of Mines publications.

Status of the Mineral Industries – Describes activities in mineral production, consumption, imports, exports and the uses of significant minerals and raw materials.

Some Bureau of Mines publications are sales publications, other series contain both free and sales publications. Sales publications may be obtained from the Superintendent of Documents, Washington, DC 20402. Free publications may be obtained from Publications Distribution Branch, Bureau of Mines, 4800 Forbes Avenue, Pittsburgh, PA 15213.

U.S. GEOLOGICAL SURVEY

Public Inquiries Office
U.S. Geological Survey
Department of Interior
19th and F Streets NW
Washington, DC 20244
Telephone: (202) 343-8073

Description: The Geological Survey maintains 10 public inquiries offices throughout the U.S. The offices provide information on the programs of the Geological Survey. Information is available on publications, open file reports, maps and data, and periodicals.

Two catalogs of Geological Survey publications are available. A monthly announcement "New Publications of the Geological Survey" is available. Public Inquiries Offices provide a referral service in order to obtain technical information and data from appropriate scientific offices.

Geologic Inquiries Group
U.S. Geological Survey
Department of Interior
907 National Center
Reston, VA 22092
Telephone: (703) 860-6517

Description: Provides information on energy and mineral resources, geologic mapping and the availability of thematic maps, geothermal energy, geophysics and geochemistry. Open filed and published reports on geologic subjects are available.

GOVERNMENT PRINTING OFFICE

GOVERNMENT PRINTING OFFICE
Superintendent of Documents
Washington, DC 20402
Telephone: (202) 783-3238 Order Desk

Description: GPO is a part of the legislative branch of the U.S. government. The Superintendent is the sales arm of GPO. The following are sources for information on publications and subscription services.

Monthly Catalog of U.S. Government Publications – Each month the library and statutory distribution service of GPO produces the monthly catalog. Entries are arranged by Superintendent of Documents classification number and contain 4 indexes: author, title, subject and series report. Complete bibliographic data are provided for each document. The Monthly Catalog is sold on a subscription basis.

Selected U.S. Government Publications – Published eleven times a year and mailed free to subscribers. Subject bibliographies of interest to the coal field are: Congressional Budget Office Publications, Energy Conservation and Resources, Federal Council for Science and Technology Publications, Government Accounting Office Publication, Grants and Awards, NASA Science and Technical Publications, NBS Publications, National Science Foundation Publications and Patents and Trademarks.

Government Periodicals and Subscription Services–Price List No. 36 – List of GPO subscriptions issued on a regular basis. The subscriptions are indexed by agency. DOE subscriptions include:

Coal Conversion Systems
Bituminous and Subbituminous Coal and Lignite
  Distribution–Energy Data Report
Coke and Coal Chemicals–Energy Data Report
Energy Abstracts for Policy Analysis
Energy Research Abstracts
Monthly Energy Review
Pennsylvania Anthracite Weekly Production
Production of Coal, Bituminous and Lignite–
  Energy Data Report

Subject Bibliographies – Publications for sale are grouped into related subjects and issued as subject bibliographies. Subjects include:

Air Pollution No. 46
Congressional Budget Office Publications No. 282
Energy Conservation and Resources No. 58
Federal Council for Science and Technology Publications No. 262
General Accounting Office Publications No. 250
Grants and Awards No. 258
Minerals Yearbooks No. 99
Minerals and Mining No. 151
National Science Foundation No. 220
Subject Bibliography Index No. 999

A complete list is available in the GPO publication "Consumer Guide to Federal Publications."

Catalog of All Publications – At present GPO does not publish a single hard copy catalog listing of the titles available for sale. A total sales catalog is published in microform. The GPO Sales Publication Reference File (PRF) is available in 48X microfiche which can be read on a microfiche reader with reduction ranging from 24X to 48X. The PRF is issued bimonthly and sold on subscription. GPO is equipped with teletype machines. Customers with teletype equipment can reach the Superintendent of Documents by dialing TWX No. 710 822-9413; answer back USGPO WSH. For further instructions, contact the local telegraph office.

Ordering Information – Orders may be charged to an established Deposit Account, Master Charge or Visa. For information call (202) 783-3238. Publications may be picked up and paid for at the GPO bookstore in Washington DC or ordered through the mail with an enclosed check or money order. For individuals in Washington who have an immediate need for publications, the Laurel, Maryland facility allows immediate pickup.

OFFICE OF TECHNOLOGY ASSESSMENT

OFFICE OF TECHNOLOGY ASSESSMENT
600 Pennsylvania Avenue SE
Washington, DC 20510
Telephone: (202) 224-8711

Description: Provides congressional committees with assessment or studies that identify the range of consequences which can be expected to accompany policy choices affecting the uses of technology. Assesses and identifies the impacts of technological programs in the areas of energy, food, genetics, population, health, materials, national security, R&D priorities and policies, oceans, technology and world trade, telecommunications, information systems and transportation.

The study "The Direct Use of Coal" was published. This study assesses the benefits and risks of such a massive shift to coal away from other fuels. It deals with the social, economic, physical and biological impacts of such a shift. It examines the complete coal system from extraction to combustion.

GENERAL ACCOUNTING OFFICE

GENERAL ACCOUNTING OFFICE
441 G Street NW
Washington, DC 20548
Telephone: (202) 275-6241

Description: GAO assists Congress and committees. Provides information, assists committees in developing statements of legislative objectives, goals and methods for assessing and reporting actual program performance, develop up to date inventory and directory of sources and information systems for fiscal, budgetary and program related information.

It publishes reports on issues studied each month. A list of these reports is available from the GAO publications office.

The Comptroller General is empowered to conduct verification examinations of energy related information developed by private business concerns. The Comptroller General may issue subpoenas, require written answers to interrogatories, administer oaths, inspect business premises and inspect copy specified books and records.

The Energy and Mineral Division is located at GAO within the Office of the Deputy Comptroller of the U.S.

Specific publications of interest to the coal industry are:

Electrical Energy Development in the Pacific Southwest—Studies fuel conservation, energy research costs, public utilities, nuclear energy, coal and utility rates.

U.S. International Energy Research and Development Program Management—Studies international cooperation, research, R&D, program management and cost sharing.

MHD: A Promising Technology for Efficiently Generating Electricity from Coal – Studies energy costs, coal facility management, electric utility construction and research programs.

Need for a System to Establish Priorities Among Fossil Energy Technologies – Studies prioritizing fossil fuels and energy research.

Long Range Planning in Department of Energy – Studies national policies, planning, systems design and evaluation and the agency's missions.

NATIONAL ACADEMY OF SCIENCES

OFFICE OF INFORMATION
National Academy of Sciences
2101 Constitution Avenue NW
Washington, DC 20418
Telephone: (202) 389-6518

Description: The Academy is a private society of individuals in scientific and engineering research. The National Academy of Engineering was established in 1964. Most of the activities undertaken are carried out through commissions and assemblies of the National Research Council.

The Commission on natural resources deals with renewable and nonrenewable resources, agriculture, energy and environmental protection. The three standing boards of the Commission are agriculture and renewable resources, environmental studies, and mineral and energy resources. The mineral and energy resource board studies energy resources and hydrocarbon, metallic and nonmetallic mineral resources and water.

Committees under the board are committees on Alaskan coal mining and reclamation, groundwater in relation to coal mining, soil as a resource in relation to surface mining for coal, mineral technology development options and chemistry of coal utilization.

Energy publications include "Technological Innovation and Forces for Change in the Mineral Industry," "Coal as an Energy Resource," and "Implications of Environmental Regulations for Energy Production and Consumption."

Coal Policy, Budget and Procurement Information

DEPARTMENT OF ENERGY

ASSISTANT SECRETARY FOR POLICY AND EVALUATION

Department of Energy
1000 Independence Avenue SW
Washington, DC 20585
Telephone: (202) 252-5325

Description: Formulates and recommends the overall national energy policy, coordinates the analysis and evaluation of policies and programs. Evaluates energy policies and programs as they apply specifically to utilities and coal supplies.

INTEGRATED PROCUREMENT MANAGEMENT INFORMATION SYSTEMS

Division of Procurement
Department of Energy
1000 Independence Avenue SW
Washington, DC 20585
Telephone: (202) 252-9061

Description: Centralized data base that collects and processes contract and procurement data. Input comes from headquarters, field offices and energy research centers.

Information given on contracts includes name of company or individuals that received contract, description of contract, city, state, award amount and completion date.

In order to obtain data on contracts, write directly to the Integrated Procurement Management Information System. The request will be programmed and costs will be assessed for the information. Information is available for a number of contract categories.

DEPARTMENT OF INTERIOR

ASSISTANT SECRETARY FOR POLICY, BUDGET AND ADMINISTRATION

Department of Interior
18th and C Streets NW
Washington, DC 20240
Telephone: (202) 343-6181

Description: Responsible for programs related to natural resources management and environmental quality, OCS program coordination, budget management, comprehensive planning, policy analysis, and economic analysis of natural and environmental resources issues. Also has authority for procurement and grants, financial and technical information systems.

OFFICE OF MINERALS POLICY AND RESEARCH ANALYSIS

Office of the Assistant Secretary for Energy and Minerals
Department of Interior
18th and C Streets NW
Washington, DC 20240
Telephone: (202) 343-8696

Description: The office is the focal point for mineral policy development and coordination within the Department of Interior.

DEPARTMENT OF COMMERCE

OFFICE OF ASSISTANT SECRETARY FOR POLICY

Department of Commerce
14th Between E and Constitution Avenue NW
Washington, DC 20230
Telephone: (202) 377-2624

Description: The Assistant Secretary advises on the development of broad departmental goals and policies with special emphasis placed on strategic resources use, development, policies and regulatory impact assessment, and domestic and international economic and business policies.

DEPARTMENT OF TRANSPORTATION

OFFICE OF THE SECRETARY FOR POLICY PLANS AND INTERNATIONAL AFFAIRS

Department of Transportation
400 7th Street SW
Washington, DC 20590
Telephone: (202) 426-4544

Description: Formulates transportation policy and plans as well as coordinating U.S. interests in international transportation affairs. Analyzes social, economic and energy aspects of transportation systems. Interfaces with local and state transportation officials.

U.S. TREASURY DEPARTMENT

UNDER SECRETARY FOR MONETARY AFFAIRS

U.S. Department of Treasury
Washington, DC 20220
Telephone: (202) 566-5164

Description: Responsible for policies in the areas of international monetary affairs, and trade and energy policies.

OFFICE OF INTERNATIONAL ENERGY POLICY

Office of the Assistant Secretary for International Affairs
U.S. Department of Treasury
Washington, DC 20220
Telephone: (202) 566-5071

Description: Deals with international monetary, commercial and energy trade policies and programs. The office deals with energy commodities and natural resources.

ASSISTANT SECRETARY FOR ECONOMIC POLICY

U.S. Department of Treasury
Washington, DC 20220
Telephone: (202) 566-2551

Description: Supervises research and analysis of major policy issues in international trade, monetary and energy areas.

ASSISTANT SECRETARY FOR TAX POLICY

U.S. Department of Treasury
Washington, DC 20220
Telephone: (202) 566-5561

Description: Provides analysis of tax programs. Also provides legal advice on domestic and international tax matters.

NATIONAL SCIENCE FOUNDATION

ENVIRONMENTAL ENERGY AND RESOURCES STUDIES

Division of Policy Research and Analyses
National Science Foundation
1800 G Street NW, Room 1240
Washington, DC 20550
Telephone: (202) 357-9689

Description: Concerned with energy policy issues as they affect nuclear, fossil fuels and other energy systems. The program studies how science and technology knowledge bases can be improved. Studies the management and use of renewable resources. Researches what flexibilities can be built into environmental strategies to allow for changes in knowledge as science and technology uncertainties are reduced.

Publishes reports that have resulted from grants and contracts. The reports are listed by subject areas, and are alphabetized by performing institutions within a given subject area.

EXECUTIVE OFFICE OF THE PRESIDENT

OFFICE OF SCIENCE AND TECHNOLOGY POLICY

Executive Office of the President
Washington, DC 20500
Telephone: (202) 456-7116

Description: Serves as a source for scientific, engineering and technological analyses, and judgement for the President with respect to major policies, plans and programs of the Federal government. Assists the President and Office of Management and Budget with the budget.

DIRECTOR FOR ENERGY, NATURAL RESOURCES AND ENVIRONMENT

White House Domestic Policy Staff
1600 Pennsylvania Avenue
Washington, DC 20500
Telephone: (202) 456-6722

Description: Involved in the administration's coal policies that include coal conversion, clean air standards, surface mining and legislation on the development of synthetic fuels.

OFFICE OF MANAGEMENT AND BUDGET

Executive Office of the President
Washington, DC 20500
Telephone: (202) 395-4840

Description: Assists the President in the preparation of the budget and formulation of the fiscal program.

Publishes *U.S. Budget in Brief, The Budget of the U.S. Government, The Budget of the U.S. Government Appendix, Special Analyses Budget of the U.S. Government* and the *Catalog of Federal Domestic Assistance.*

Incorporated within OMB is the Office of Natural Resources Energy and Science. This office is involved in the preparation of the budget incorporating the evaluations of OMB science programs.

The Office of Federal Procurement Policy provides overall direction of procurement policies, regulations procedures and forms.

CONGRESSIONAL BUDGET OFFICE

CONGRESSIONAL BUDGET OFFICE

2nd and D Streets SW
Washington, DC 20515
Telephone: (202) 225-4416

Description: Provides Congress with basic budget data and with analyses of fiscal, budgetary and programmatic policy issues. CBO provides periodic forecasts and analysis of economic trends and alternative fiscal policies. CBO monitors actions on individual authorizations, appropriations and revenue bills against the targets or ceilings specified in the concurrent resolutions. Provides 5-year projections on the costs of continuing Federal spending and taxation policies. Presents an annual report by April 1 of each year to the House and Senate committees. Performs special studies in budget-related areas. CBO has a publication entitled "Replacing Oil and Natural Gas with Coal: Prospects in the Manufacturing Industries" August 1978.

Coal-Related Legislative Committees and Advisory Committees

SENATE

ARMED SERVICES COMMITTEE
212 Russell Senate Office Building
Washington, DC 20510
Telephone: (202) 224-3871

Description: Jurisdiction over energy supplies and conservation for the military.

COMMITTEE ON APPROPRIATIONS
Dirksen Senate Office Building
Washington, DC 20510
Telephone: (202) 224-3471

Description: Jurisdiction over legislation on appropriations and new spending authority.

Coal-Related Subcommittees –

Subcommittee on Energy and Water Development
Subcommittee on HUD and Independent Agencies
Subcommittee on Interior
Subcommittee on Labor, Health, Education & Welfare

COMMITTEE ON BANKING, HOUSING AND URBAN AFFAIRS
5300 Dirksen Senate Office Building
Washington, DC 20510
Telephone: (202) 224-7391

Description: Jurisdiction over legislation on export and export controls as they pertain to coal.

Coal-Related Subcommittee –

Subcommittee on Housing and Urban Affairs

COMMITTEE ON THE BUDGET
Carroll Arms Hotel
Washington, DC 20510
Telephone: (202) 224-0642

Description: Makes continuing studies of the effects on budget outlays of relevant existing and proposed legislation.

COMMITTEE ON COMMERCE, SCIENCE AND TRANSPORTATION
5202 Dirksen Senate Office Building
Washington, DC 20510
Telephone: (202) 224-5115

Description: Jurisdiction over legislation concerning science, engineering, technology, research and development and policy.

Coal-Related Subcommittees –

Subcommittee on Science Technology and Space
Subcommittee on Surface Transportation

COMMITTEE ON ENERGY AND NATURAL RESOURCES
3106 Dirksen Senate Office Building
Washington, DC 20510
Telephone: (202) 224-4971

Description: Jurisdiction over coal slurry pipelines, coal policy, regulations, conservation, R&D, extraction of minerals from oceans, hydroelectric power, irrigation and reclamation, mining education, research mining, mineral lands, mining claims and public lands, forests and energy resource development.

Coal-Related Subcommittees –

Subcommittee on Energy Conservation and Supply
Subcommittee on Energy Regulation
Subcommittee on Energy R&D
Subcommittee on Energy Resources and Materials Production
Subcommittee on Parks Recreation and Renewable Resources
Subcommittee on Energy Conservation and Supply

COMMITTEE ON ENVIRONMENT AND PUBLIC WORKS
4204 Dirksen Senate Office Building
Washington, DC 20510
Telephone: (202) 224-6176

Description: Concerned with air pollution, environmental policy and environmental R&D.

Coal-Related Subcommittees –

Subcommittee on Environmental Pollution
Subcommittee on Resource Protection
Subcommittee on Regulations & Community Development

COMMITTEE ON FINANCE
227 Dirksen Senate Office Building
Washington, DC 20510
Telephone: (202) 224-4515

Description: Jurisdiction over the transportation of goods for energy-related industries.

Coal-Related Subcommittees –

Subcommittee on Energy & Foundations
Subcommittee on Tax and Debt Management

COMMITTEE ON FOREIGN RELATIONS
4229 Dirksen Senate Office Building
Washington, DC 20510
Telephone: (202) 224-4651

Description: Jurisdiction over the international aspects of energy-related policies.

Coal-Related Subcommittee –

Subcommittee on International Economic Policy

COMMITTEE ON GOVERNMENTAL AFFAIRS
3308 Dirksen Senate Office Building
Washington, DC 20510
Telephone: (202) 224-4751

Description: Involved with agencies that deal with energy statistics, shortages and energy programs.

Coal-Related Subcommittees –

Subcommittee on Energy Nuclear Proliferation
Subcommittee on Investigations

COMMITTEE ON LABOR AND HUMAN RESOURCES
4230 Dirksen Senate Office Building
Washington, DC 20510
Telephone: (202) 224-5375

Description: Concerned with occupational safety and health. Includes the welfare of miners.

Coal-Related Subcommittees –

Subcommittee on Health & Science
Subcommittee on Taxation Financing & Investment

HOUSE

ARMED SERVICES COMMITTEE
2120 Rayburn House Office Building
Washington, DC 20515
Telephone: (202) 225-4151

Description: Jurisdiction over energy reserves and supplies for the military.

COMMITTEE ON APPROPRIATIONS
Capitol H218
Washington, DC 20515
Telephone: (202) 225-2771

Description: Jurisdiction over appropriations for energy programs within the government agencies.

Coal-Related Subcommittees –

Subcommittee on Energy & Water Development
Subcommittee on State, Justice, Commerce and Judiciary
Subcommittee on Labor, Health, Education & Welfare
Subcommittee on Transportation

COMMITTEE ON BANKING, FINANCE AND URBAN AFFAIRS
2129 Rayburn House Office Building
Washington, DC 20515
Telephone: (202) 225-4247

Description: Areas of interest include financial aid to commerce and industry. Studies financial incentives in the use of energy conservation within industries.

Coal-Related Subcommittees –

Subcommittee on Economic Stabilization
Subcommittee on International Investigation and Monetary Policy

COMMITTEE ON THE BUDGET
A214 HOB Annex 1
Washington, DC 20515
Telephone: (202) 225-7200

Description: Concerned with fiscal policy and budget recommendations.

COMMITTEE ON EDUCATION AND LABOR
2181 Rayburn House Office Building
Washington, DC 20515
Telephone: (202) 225-4527

Description: Concerned with mediation and arbitration of labor disputes.

Coal-Related Subcommittee –

Subcommittee on Health and Safety

COMMITTEE ON FOREIGN AFFAIRS
2170 Rayburn House Office Building
Washington, DC 20515
Telephone: (202) 225-5021

Description: Jurisdiction over energy problems that involve U.S. foreign policy.

Coal-Related Subcommittees –

Subcommittee on International Economic Policy & Trade
Subcommittee on International Security and Scientific Affairs

COMMITTEE ON GOVERNMENT OPERATIONS
2154 Rayburn House Office Building
Washington, DC 20515
Telephone: (202) 225-5051

Description: Involved in energy matters involving energy reorganization issues within energy-related agencies.

Coal-Related Subcommittee –

Subcommittee on Environment, Energy and Natural Resources

COMMITTEE ON INTERIOR AND INSULAR AFFAIRS

1324 Longworth House Office Building
Washington, DC 20515
Telephone: (202) 225-2761

Description: Jurisdiction over mineral land laws and claims, mining interests, mining schools and experimental stations. Exercises jurisdiction over environmental issues related to energy.

Coal-Related Subcommittees –

Subcommittee on Mines and Mining
Subcommittee on Public Lands
Subcommittee on Energy and Environment
Subcommittee on Water and Power Resources

COMMITTEE ON INTERSTATE AND FOREIGN COMMERCE

2125 Rayburn House Office Building
Washington, DC 20515
Telephone: (202) 225-2927

Description: Jurisdiction over regulation of interstate transmission of power. Concerned with energy policy and clean air legislation.

Coal-Related Subcommittees –

Subcommittee on Energy & Power
Subcommittee on Health & Environment
Subcommittee on Transportation & Commerce

COMMITTEE ON PUBLIC WORKS AND TRANSPORTATION

2165 Rayburn House Office Building
Washington, D.C. 20515
Telephone: (202) 225-4472

Description: Jurisdiction over energy matters which relate to public buildings.

Coal-Related Subcommittee –

Subcommittee on Surface Transportation

COMMITTEE ON SCIENCE AND TECHNOLOGY

2321 Rayburn House Office Building
Washington, DC 20515
Telephone: (202) 225-6371

Description: Jurisdiction over all energy R&D as well as demonstration.

Coal-Related Subcommittees –

Subcommittee on Energy Development & Application

Subcommittee on Energy Research and Production
Subcommittee on Natural Resources and Environment
Subcommittee on Science Research and Technology

COMMITTEE ON SMALL BUSINESS
2361 Rayburn House Office Building
Washington, DC 20515
Telephone: (202) 225-5821

Description: Jurisdiction over energy programs for small business. Concerned with financial incentives.

Coal-Related Subcommittees –

Subcommittee on Impact of Energy Programs, Environment and Safety Requirements and Government Research on Small Business
Subcommittee on Energy, Environment, Safety & Research

COMMITTEE ON WAYS AND MEANS
1102 Longworth House Office Building
Washington, DC 20515
Telephone: (202) 225-3625

Description: Jurisdiction over tax incentives for energy activities and programs.

Coal-Related Subcommittees –

Subcommittee on Health
Subcommittee on Trade

JOINT COMMITTEE

JOINT ECONOMIC COMMITTEE
G 133 Dirksen Senate Office Building
Washington, DC 20510
Telephone: (202) 224-5171

Description: Holds hearings on budget proposals including energy matters.

Coal-Related Subcommittees –

Subcommittee on Energy
Subcommittee on Fiscal and Intergovernmental Policy
Subcommittee on Priorities and Economics in Government
Subcommittee on Economic Growth and Stabilization

ADVISORY COMMITTEES

COMMITTEE MANAGEMENT SECRETARIAT
General Services Administration
Washington, DC 20405
Telephone: (202) 357-0019

Description: Advisory committees range from providing policy advice on major national issues to providing specific technical recommendations on a particular problem. Industry and government officials participate on panels. These meetings are announced in the Federal Register and copies of meeting transcripts and lists of members are available. Contact the committee management officer or staff contact person in each agency responsible for the committee or commission.

Coal-Related Committees and Commissions –

Advisory Committee for Engineering–NSF
Advisory Committee for Policy Research & Analysis and Science Resources Studies–NSF
Advisory Committee on Mining and Mineral Resources Research–Department of Interior
Advisory Committee on Science, Technology and Development–Office of Science Technology and Policy
Clean Air Scientific Advisory Committee–EPA
Coal Industry Advisory Committee–DOE
Energy Research Advisory Board–DOE
Environmental Advisory Committee–DOE
Federal Advisory Council on Occupational Safety and Health–Department of Labor
Fossil Energy Advisory Committee–DOE
Industry Sector Advisory Committee on Construction, Mining, Agricultural and Oil Field Machinery and Equipment for Multilateral Trade Negotiations–Department of Commerce
Mine Health Research Advisory Committee–HHS
National Advisory Committee on Occupational Safety and Health–Department of Labor
National Advisory Environmental Health Sciences Council–HHS
National Air Pollution Control Techniques Advisory Commission–EPA
National Industrial Energy Council–DOE
National Transportation Policy Study Commission
President's Commission on Coal Industry
Commission on an Agenda for the Eighties

Legal and Regulatory Data

DEPARTMENT OF ENERGY

OFFICE OF THE SECRETARY
Department of Energy
1000 Independence Avenue SW
Washington, DC 20585
Telephone: (202) 252-6210

Description: The Office of the General Counsel, Office of Hearings and Appeals, Economic Regulatory Administration, and The Federal Energy Regulatory Commission report to the Secretary's Office.

Office of the General Counsel — Provides legal counsel to DOE on legislative activities, litigation, advice on international cooperation agreements and the transfer of energy-efficient products and technologies from the research stage to commercial use.

Office of Hearings and Appeals — Cases are listed in the Federal Register received by this office. Information is listed by date, name and location of applicant, case number, type of submission and notices of objections received.

Economic Regulatory Administration — Administers the department's regulatory programs other than those assigned to the FERC. Programs include conversion of oil- and gas-fired utility and industrial facilities to coal, natural gas, import/export controls, natural gas curtailment priorities and emergency allocations, regional coordination of electric power system planning and reliability of bulk power supply.

Federal Energy Regulatory Commission — Independent commission within DOE that sets rates and charges for transportation and sale of natural gas and sale of electricity and the licensing of hydroelectric power projects.

DEPARTMENT OF INTERIOR

OFFICE OF THE SECRETARY
Department of Interior
18th and C Streets NW
Washington, DC 20240
Telephone: (202) 343-7351

Description: Several departments concerned with legal matters pertaining to all aspects of the coal industry report to the Secretary's office.

Office of the Solicitor — Performs all the legal work of the Department with the exception of that performed by the Office of Hearings and Appeals and the Office of Congressional and Legislative Affairs. The Office is divided into several divisions concerned with coal matters.

The Division of Energy and Resources — Responsible for legal matters involving the programs of the Assistant Secretary for Energy and Minerals, Bureau of Land Management, Bureau of Mines, and the Geological Survey.

Division of Surface Mining — Provides advice on surface mining matters and cases to the Office of Surface Mining Reclamation and Enforcement.

Division of General Law — Responsible for general administrative law matters and legal matters involving programs under the jurisdiction of the Assistant Secretary for Policy Budget and Administration.

Office of Hearings and Appeals — Responsible for departmental quasi-judicial and related functions. Renders decisions on public and acquired lands and their resources, surface coal mining control and reclamation.

DEPARTMENT OF JUSTICE

LAND AND NATURAL RESOURCES DIVISION,
ASSOCIATE ATTORNEY GENERAL
Department of Justice
Constitution Avenue and 10th Street NW
Washington, DC 20530
Telephone: (202) 633-2701

Description: Responsible for the conduct of lawsuits both in federal and state courts relating to the protection of interests in real property and natural resources owned or sought to be acquired by the federal government.

Establishes rights in minerals and other natural resources. Establishes rights to abate water, air and noise pollution. Responsible for criminal prosecutions for air, water and noise pollution.

Defends officers of the U.S. with respect to their actions relating to federal lands and resources. The Division defends suits against government officers arising out of the National Environmental Policy Act.

ANTITRUST DIVISION
Department of Justice
Constitution Avenue and 10th Street NW
Washington, DC 20530
Telephone: (202) 633-2401

Description: The publication produced by the division entitled "Competition on Coal Industry" analyzes the antitrust provisions of the Mineral Leasing Act. Determines whether antitrust laws are viable in providing competition in the coal industry.

INTERSTATE COMMERCE COMMISSION

RATES SECTION
Interstate Commerce Commission
12th and Constitution Avenue NW
Washington, DC 20423
Telephone: (202) 275-7693

Description: Regulates the movement and rates of coal by rail and pipelines. Involved in providing adequate freight space for coal. The ICC document room has information on cases, regulations, and hearings pertaining to matters that deal with all the means of transportation that affect the coal industry.

SECURITIES AND EXCHANGE COMMISSION

PUBLIC AFFAIRS
Securities and Exchange Commission
500 N Capitol Street
Washington, DC 20549
Telephone: (202) 272-2650

Description: Regulates, reviews and discloses information relating to public companies involved in the coal industry.

FEDERAL TRADE COMMISSION

ECONOMIC BUREAU
Federal Trade Commission
2120 L Street NW
Washington, DC 20580
Telephone: (202) 254-7720

Description: Researches data on industrial and business matters pertaining to the coal industy. Studies and supports legislation to provide competition in the coal industry marketplace.

Coal Research and Technology

DEPARTMENT OF ENERGY

ASSISTANT SECRETARY FOR RESOURCE APPLICATIONS
Department of Energy
12th and Pennsylvania Avenue NW
Washington, DC 20461
Telephone: (202) 623 9222

Description: The Office of Industrial and Utility Applications and Operations within the Assistant Secretary's office promotes commercial application of high Btu coal gasification, low/med Btu coal gasification, coal liquids, coal conversion, coal production, coal technology and coal supplies.

ASSISTANT SECRETARY FOR FOSSIL ENERGY
Department of Energy
1000 Independence Avenue SW
Washington, DC 20585
Telephone: (301) 353 2642

Description: This office studies coal technology, utilization, processing and coal fired MHD systems. The Director of Advanced Research and Technology with the Assistant Secretary's Office develops mid and long term energy technology and development strategies.

OFFICE OF ENERGY RESEARCH
Department of Energy
1000 Independence Avenue SW
Washington, DC 20585
Telephone: (202) 252-5430

Description: Advises the Secretary on all the department's R&D programs. Manages basic energy sciences programs, administers DOE programs supporting university research, and administers financial support for R&D projects not funded elsewhere in the department.

Coal R&D is conducted at the following DOE field installations: Ames Laboratory (coal chemistry); Argonne National Laboratory (coal conversion, high sulfur coal); Bartlesville Energy Technology Center (coal conversion, synthetic fuels); Carbondale Mining Operations Office (mining research); Grand Forks Technology Center (coal combustion, gasification, and liquefaction); Idaho Operations Office (MHD); Laboratory for Energy Related Health (coal environmental research); Laramie Energy Technology Center (coal gasification, in situ recovery techniques); Lawrence Berkeley Laboratory (fossil fuels); Lawrence Livermore Laboratory (coal gasification); Morgantown Energy Technology Center (coal gasification, combustion); Oak Ridge Operations Office (coal liquefaction); Pacific Northwest Laboratory (MHD, coal gasification); Pittsburgh Energy Technology Center (coal combustion, conversion, MHD); Savannah River Operations Office (coal environmental research).

ENERGY RESEARCH, DEVELOPMENT AND DEMONSTRATION INVENTORY
Department of Energy
Information Center Complex/Information Division
Oak Ridge National Laboratory
Bldg. 3603
P.O. Box X
Oak Ridge, TN 37830
Telephone: (615) 574-7803

Description: Contains a computerized inventory containing descriptions of current energy related research. The scope of interest includes all energy sources. Information is available on coal exploration mining and resources. The descriptions of energy research are arranged by subject categories and consist of when available: title, research institution and city, sponsor, principal investigator, project duration, funding level, description of research, number of technical personnel assigned to the project, type of research and publications. Publishes "Inventory of Energy Research and Development."

TECHNICAL INFORMATION CENTER
Department of Energy
P.O. Box 62
Oak Ridge, TN 37830
Telephone: (615) 576-1188

Description: A program guide describing DOE research needs and other opportunities is available in a publication entitled "The DOE Program Guide for Universities and Other Research Groups." The publication presents information on each major DOE R&D program including program objectives, problem areas requiring additional research, potential areas for new research initiatives, program information contacts and total R&D budget data for FY 1980 and 1981. Part II of the guide describes existing Federal assistance and procurement policies and procedures, and discusses those items currently under development or revision that will affect universities or other research organizations associated with DOE. Information is included pertaining to the submission and evaluation of proposals, and the administration of resulting grants, cooperative agreements and research contracts. A chart depicting the DOE field organization, with names, addresses and telephone numbers of key program contacts is included. The publication is free.

DEPARTMENT OF INTERIOR

ASSISTANT SECRETARY FOR ENERGY AND MINERALS

Department of Interior
18th and C Streets NW
Washington, DC 20240
Telephone: (202) 343-2186

Description: Responsible for programs associated with mineral policy, data and analysis, surface mining reclamation, and mining research and development. Coal research related offices and departments within the Office of Energy and Minerals:

The Office of Minerals Policy and Research Analysis – Oversees the development of new minerals research and development programs. Evaluates the progress and results of all minerals R&D programs within the Department as well as advises the Assistant Secretary for Energy and Minerals on the development of mineral programs and associated research programs.

The Bureau of Mines – Responsible for research to develop the technology for the extraction, processing, use and recycling of mineral resources. Areas of research include mine health and safety, recycling of solid wastes, and abatement of pollution and land damages caused by mineral extraction.

Coal research related offices within the Bureau of Mines:

The Minerals Research Directorate has jurisdiction over the Division of Minerals Environmental Technology, Division of Minerals Health and Safety Technology, Division of Minerals Resource Technology, Division of Research Center Operations and the Division of Helium Operations.

The Divisions have the following responsibilities:

Division of Minerals Environmental Technology has responsibility for environmental assessment. This branch is the focal point for research, studies, and public works activities concerning abandoned mine lands and environmental problems arising from past and present mining and mineral processing. Areas of interest include mined land reclamation, subsidence control, physical and vegetative stabilization of waste banks, extinguishment or control of underground outcrop and mine refuse, bank fires, sealing oil and gas wells, pollution control for abandoned underground mines, recovery of anthracite coal from refuse banks and reclamation of surface mined land for a variety of public uses.

Published information circulars, special investigation reports, and contractor reports. Answers inquiries, provides advisory services to public bodies, makes referrals and provides information on program and project work in progress.

Division of Minerals Health and Safety has responsibility for health and safety research and technology transfer.

Division of Minerals Resource Technology has responsibility for mineral science and technology, resource conservation and resource development.

Division of Research Center Operations has responsibility for the following centers: Avondale Research Center, Twin Cities Research Center, Salt Lake City Research Center, Rolla Research Center, Reno Research Center, Albany Research Center, Tuscaloosa Research Center, Pittsburgh Research Center, Spokane Research Center and Denver Research Center. The Mining Research Center at Denver, Colorado areas of interest include rock mechanics as applied to ground control stress, strength relationships in mine openings, and design of adequate safe openings for underground and open pits. Research is conducted on the safety aspects of coal and metal mining. Reports and studies are available. Answers inquiries and makes referrals.

The Coal Sampling Inspection Office in College Park, Maryland, is partially sponsored by other federal government agencies and the Bureau of Mines. It issues coal analyses reports to federal agencies on coal purchased under contract. Issues tipple analysis requests to coal operators. Provides technical and consulting services to government agencies and offices throughout the U.S. on problems relating to methods of sampling and the purchase, storage, use, availability and inspection of coal. Areas of interest include chemical analysis of coal, coal constituents, fossil fuels, lignite, carbonaceous rocks and heating fuels.

Publishes "Analyses of Coal" – Annual Report, and "Analyses of Tipple and Delivered Samples of Coals." Answers inquiries, provides advisory, reference literature searching, abstracting, indexing, current awareness services, provides information on research in progress, distributes data compilations and publications. Permits onsite use of computerized information. Holdings include computerized information for the "Analyses of Tipple and Delivered Samples of Coal." Computerized data are available on current and historical analyses of coal.

The U.S. Geological Survey – Performs surveys, investigations, and research covering topography, geology, and the mineral and water resources of the U.S. Also classifies land as to mineral character and water and power resources. Enforces departmental regulations applicable to oil, gas and other mining leases, permits, licenses, development contracts, and gas storage contracts.

Coal research related division within the U.S. Geological Survey:

The Geologic Division conducts research programs to aid in the management of energy potential in U.S. These programs provide basic information on the character, magnitude, location and distribution of mineral, energy, and land resources as well as on the principles and processes involved in their formation. The division is involved in mineral land assessments.

DEPARTMENT OF DEFENSE

OFFICE OF THE DIRECTOR FOR RESEARCH

Research and Advanced Technology
Department of Defense
Pentagon
Washington, DC 20301
Telephone: (202) 697-3228

Description: The objective of the energy program is to reduce DOD's activities on foreign oil through the future use of domestic synthetic fuel, conservation and the use of other fuels and energy sources.

The Defense department is developing new engines capable of using a broad range of fuels. DOD is evaluating several liquid hydrocarbon fuels derived from low quality petroleum crudes, oil shale and coal for use in military turbine engines. The major thrusts of the DOD synfuels program are directed toward the application and the development of specific technologies that will encourage

DOD to develop synfuels for many areas of military use, and to develop self-sufficiency on military installations.

DEPARTMENT OF COMMERCE

ASSISTANT SECRETARY FOR PRODUCTIVITY, TECHNOLOGY AND INNOVATION
U.S. Department of Commerce
14th between Constitution and E Streets NW
Washington, DC 20230
Telephone: (202) 377-3111

Description: Involved in R&D activities pertaining to science and technology programs as well as the department's environmental impact statements. Oversees NBS, NTIS, Patent and Trademark Office, and Office of Environmental Affairs.

NATIONAL BUREAU OF STANDARDS
Department of Commerce
Route 270
Germantown, MD 20234
Telephone: (301) 921 3112

Description: Responsible for standards and measurements for synthetic and fossil fuels. Specific offices with energy and environmental data:

National Engineering Laboratory – Conducts research in engineering and applied sciences. Provides engineering measurement traceability services. Develops, tests and proposes new energy practices and developments. Improves mechanisms to transfer results of its research to users.

National Standard Reference Data System – NSRDS has the mission of providing critically evaluated numeric data in a convenient and accessible form to the scientific and technical community. Projects fall into the categories of energy and environmental data, industrial process data, materials utilization data, and physical science data. Administers a number of data analysis centers.

ENVIRONMENTAL PROTECTION AGENCY

OFFICE OF THE ASSISTANT ADMINISTRATOR FOR R&D

Environmental Protection Agency
401 M Street SW
Washington, DC 20460
Telephone: (202) 755-2600

Description: Researches methods to control and abate adverse impacts on the environment associated with the development and utilization of energy systems and nonrenewable resources. Functions include management of selected demonstration programs and planning for agency environmental quality monitoring programs.

The Office of R&D supervises the Industrial Research Laboratories. The Laboratory at Research Triangle Park in North Carolina researches combustion technology, methods for the disposal of effluents from power plants and provides an assessment of environmentally related information on coal gasification and liquefaction.

Coal research related office within the Office of Research and Development:

> *The Office for Energy, Minerals and Industry* – Involved in research in the areas of coal cleaning, extraction, combustion, synfuels, advanced systems and the environmental effects of coal.

NATIONAL AERONAUTICS AND SPACE ADMINISTRATION

OFFICE OF PUBLIC AFFAIRS

NASA
400 Maryland Avenue SW
Washington, DC 20546
Telephone: (202) 755-3828

Description: The following programs and field installations have data of value pertaining to energy technology, resources and environmental information.

> *Office of Aeronautics and Space Technology* – Identifies technologies developed in the space program that have the potential for making major contributions to the solution of energy problems and to provide technical support to other agencies in developing specific applications.

Space and Terrestrial Applications – Responsible for resource observation.

Space Science – Researches the earth, solar system and the universe.

Lyndon B. Johnson Space Center – Involved in earth resources surveys.

Langley Research Center – Involved in environmental quality monitoring.

National Space Technology Laboratories – Houses certain environmental research and earth resources activities of NASA.

The following are the informational sources and systems that provide data on research and technology.

Research and Technology Objectives and Plans Summary – This document lists supporting research and technology currently in process at NASA. Contains abstracts from approved research and technology objectives and plans. Available from NTIS.

Scientific and Technical Information System – Supports the NASA R&D efforts and assists in the dissemination of NASA program generated information to the public. The system covers all the sciences and technologies related to aeronautics, space-earth resource surveys, solar and wind energy and other areas of NASA R&D. Also received are publications issued by DOD, FAA, DOE, EPA, NOAA, NSF and NTIS. Scientific and Technical Aerospace Reports are available. Subjects include science and technology as related to space. STAR announces reports, doctoral theses, NASA patents and includes current research projects. Publications from internal sources issued through STAR include Contractors Reports, Technical Memoranda, and Technical Notes and Reports. Available from GPO.

Periodic Bibliographies – Subject areas include aeronautical engineering, aerospace medicine and biology, computer program abstracts, earth resources, energy, management, NASA patent abstracts and high energy propellants. The bibliography on energy resources includes environmental changes and geology, and mineral resources. The bibliography on energy includes primary energy sources and their secondary sources, energy conversion and energy storage.

Office of Technology Utilization – Each NASA Center has a Technology Utilization Officer. The Office of Technology

Utilization publishes "Tech Briefs." The publication describes potential products, industrial processes, basic and applied research, computer software and new sources of technical data.

Industrial Application Centers – Maintain computerized access to space related reports and reports from onspace governmental sources. Major information sources include Air Pollution Technical Information Center, Chemical Abstracts Condensates, Engineering Index, Government Reports Announcements, NASA International Aerospace Abstracts, NASA Tech Briefs, NASA's STAR, Nuclear Science Abstracts, Selected Water Resources Abstracts, and Educational Resources Information Center.

The University of New Mexico, Institute for Applied Research Services, located in Albuquerque, New Mexico has information on biology and chemistry, business and government, education and social sciences, engineering, environment and pollution, geology and mineral resources, and energy conservation. The data is used by this center to operate its major information programs in industrial information, energy information and remote sensing natural resources information.

CONSERVATION AND FOSSIL ENERGY SYSTEMS
NASA
400 Maryland Avenue SW
Washington, DC 20546
Telephone: (202) 755-8475

Description: NASA's primary interest in fossil fuels involves research into advanced mining techniques, cleansing of coal prior to combustion and conservation. Information is available through NASA's publications, information systems, technical notes and contractor reports.

NATIONAL SCIENCE FOUNDATION

EARTH SCIENCES, DIRECTORATE FOR ASTRONOMICAL, ATMOSPHERIC, EARTH AND OCEAN SCIENCES
National Science Foundation
1800 G Street NW
Washington, DC 20550
Telephone: (202) 357-7915

Description: The earth sciences programs are devoted to increasing basic knowledge of the earth and its ocean floor. Earth Sciences supports research to obtain basic information about the nature, origin, history and behavior of rock formations. This enables researchers to understand the processes that have produced petroleum, coal, metals, and other earth resources.

Major research areas of NSF programs in Earth Sciences are:

Geology – Field and laboratory research on such geological processes as volcanic eruption, movement of glaciers, and erosion.

Geochemistry – Studies the composition, age, and origin and development of rocks and minerals. Researches the processes involved in the transportation and segregation of chemical elements in the earth's crust.

Geophysics – Studies the physical state and properties of the earth. Subfields covered are seismology, gravity, geodesy, rock magnetism, terrestrial currents, heat flow, and high pressure phenomena.

DIRECTORATE FOR SCIENTIFIC, TECHNOLOGICAL AND INTERNATIONAL AFFAIRS
National Science Foundation
1800 G Street NW
Washington, DC 20550
Telephone: (202) 357-7631

Description: The Environmental Energy and Resources Studies program studies linkages among energy, environmental constraints and the management and use of renewable resources. Studies the interaction between alternative energy options and technology public policy decisions. The Division of Science Resources Studies provides a central clearinghouse for the collection, interpretation, and analysis of data on the availability of, and the current and projected needs for scientific and technical resources in the United States.

DIRECTORATE FOR MATHEMATICS AND PHYSICAL SCIENCES
National Science Foundation
1800 G Street NW
Washington, DC 20550
Telephone: (202) 357-9742

Description: Support is given to research in the areas of chemistry, materials research, computer sciences, mathematical sciences, and physics.

INTEGRATED BASIC RESEARCH, DIRECTORATE FOR ENGINEERING AND APPLIED SCIENCE
National Science Foundation
1800 G Street NW
Washington, DC 20550
Telephone: (202) 357-9666

Description: Supports research in order to obtain scientific knowledge for future technologies in nonfuel mineral exploration and mining. This includes knowledge leading to a deep solution mining capability through basic research in the chemistry of mineral leaching, rock fracture mechanics, reservoir engineering of chemically active fluids, and geotechnical sensing of subsurface chemical and flow regimes. Research will enable a basic understanding of the role of minerals on earth.

TENNESSEE VALLEY AUTHORITY

TENNESSEE VALLEY AUTHORITY
400 Commerce Street
Knoxville, TN 37902
Telephone: (615) 632-3257

Description: TVA has programs to demonstrate promising technologies for better use of coal. These programs are fuel cells using gas from coal, fluidized bed combustion, and advanced sulfur dioxide control processes for conventional fired boilers. TVA is participating in the organization of a new national electric utility program to provide financial support for power research and development.

Patents, Inventions and Technology Transfer Sources

DEPARTMENT OF COMMERCE

U.S. PATENT AND TRADEMARK OFFICE

Department of Commerce
2021 Jefferson Davis Highway
Arlington, VA 20231
Telephone: (703) 557-3158

Description: Publishes and disseminates information including copies of patents. The Patent Search files provide a collection of more than 4 million distinct U.S. patents. These patents are classified and cross-referenced among approximately 100,000 categories of technology that make up the U.S. Patent Classification System. The files are used by patent examiners, patent attorneys and inventors in search of prior information in relation to filing or prosecuting patent applications, by individuals seeking a specific patent and by the general public.

Publications available through GPO include: General Information Concerning Patents, Annual Indexes, Patent Official Gazette, Manual of Patent Examining Procedures, Manual Classification of Patents and Inventions. The Search room is open to the public. Copies of the specifications and drawings on all patents are available.

Specific sources within the Patent and Trademark Office include:

> *Office of Technology Assessment and Forecast* – The purpose of OTAF is to assemble, analyze and make available data and to identify clear trends in patenting and areas of technology in which there is a high proportion of activity. Provides business and government with a single source from which to obtain in-

formation covering the entire spectrum of technology. Special areas include patents and patent data, technology assessment for technology, energy and energy technology and patent ownership. OTAF has built a master data base covering all U.S. patents.

OTAF issues general distribution publications. The technology assessment and forecast reports review highly active technological areas on U.S. and foreign patents. These reports are available from NTIS. The third report, June 1974 COM 74-11383, presents an overview of the technological activity, across all technologies of a group of selected U.S. states. Extends energy area treatment to include oil shale and coal gasification technologies.

The fourth report, January 1975 COM 75-10050, reviews in terms of patent activity during the years 1969-1973, nuclear energy, oil shale and coal gasification technologies.

The eighth report, December 1977 PB 276375, reviews U.S. patenting in the context of domestic versus international patenting and analyzes the balance of patenting between U.S. and other countries. Concludes with an in-depth analysis of patent activity in geophysical exploration for hydrocarbons.

OTAF has a series of publications called Patent Profiles Surveys. Describes patenting activities in specific technologies. The first two issues are directed to synthetic fuels and solar energy.

OTAF prepares special reports tailored to individual needs. They assemble, analyze and make available in a number of formats meaningful information about the patent files. All special reports are prepared on a cost reimbursable basis. These costs may vary from as low as $35.00 for some forms to several thousand dollars for complex and large-scale treatments of many technological categories.

Scientific Library -- Contains over 120,000 volumes of scientific and technical books. Information is provided on foreign patents.

NATIONAL TECHNICAL INFORMATION SERVICE
U.S. Department of Commerce
5285 Port Royal Road
Springfield, VA 22161
Telephone: (703) 487-4650

Description: NTIS is the central source for information for all new U.S. Government owned patents and patent pending applications. These inventions primarily come out of government laboratories but also include contractor inventions to which the government has title. Patents and inventions are announced in the following publications:

Government Inventions for Licensing – Inventions are summarized in this newsletter when government agencies submit new inventions to NTIS and when patent applications are filed with the U.S. Patent and Trademark Office and again when patents are issued.

Patent Portfolio – Inventions are cited and most of these are available for licensing. Consists of 3 subject volumes (Chemical, Electrical, and General/Mechanical) and an index with key word and number references.

Selected Technology for Licensing – The most commercially promising inventions are summarized. The publication is available through the NTIS "Tech Notes."

NASA Patent Abstracts Bibliography – The abstract section has six months of annotated references to NASA-owned inventions and applications for patents announced in Scientific and Technical Aerospace Reports between January-June and July-December. Each index section is cumulative and covers entries for all patent and patent application citations since May 1969 and is also issued semiannually.

DEPARTMENT OF ENERGY

U.S. DOE PATENTS AVAILABLE FOR LICENSING

Assistant General Counsel for Patents
Department of Energy
Washington, DC 20545
Telephone: (301) 353-4970

Description: Nonexclusive royalty-free revocable licenses are granted by DOE upon request to individuals and corporations. Exclusive and limited exclusive licenses may also be granted.

Applications may be made by letter with information on: (1) type of applicants' business, (2) purpose for which the license is desired, (3) geographical areas in which the applicant will practice the invention, (4) applicants' willingness to render periodic reports on the use of the licensed subject matter, (5) applicants' status as a small or minority business, and (6) a remittance of $10.00. A quarterly report is published on DOE-owned U.S. patents and patent applications available for licensing.

ENERGY RELATED INVENTIONS
Department of Energy-National Bureau of Standards
Route 270
Gaithersburg, MD 20234
Telephone: (301) 921-3694

Description: Assists inventors with their ideas to develop new non-nuclear energy related technology. Cooperates with NBS's Office of Energy Related Inventions Program. If an invention is recommended, it may receive financial support. This support may be given by means of an unsolicited proposal. Inventors should request Energy Related Inventions Evaluation Form.

TECHNOLOGY TRANSFER, PLANS AND TECHNOLOGY ASSESSMENT
Fossil Energy
Department of Energy
Washington, DC 20545
Telephone: (301) 353-2782

Description: Programs are aimed at developing technology for transfer to industry. The goals are to develop processes to the commercial scale for producing clean fuels from coal, to increase utilization of coal as a primary fuel. Technology transfer's responsibility is a major management function of all technical areas and the responsibility lies with the program managers.

DEPARTMENT OF INTERIOR

PATENTS BRANCH
Department of Interior
18th and C Streets NW
Washington, DC 20240
Telephone: (202) 343-4471

Description: Studies patents from department employees and contractors. Interested in patents if they pertain to mineral research and exploration, mine safety and pollution control.

TECHNOLOGY TRANSFER GROUP
Bureau of Mines
Department of Interior
2410 E Street NW
Washington, DC 20241
Telephone: (202) 634-1225

Description: The group informs industry, mining personnel and educators of the results of Bureau funded research. Conducts briefings, seminars, exhibits, demonstrations. Provides cost sharing contracts, cooperative agreements and information on patent licensing. Publishes "Technology News," "Mining Research Review" and "Mining Technology Research."

ENVIRONMENTAL PROTECTION AGENCY

TECHNICAL INFORMATION OPERATIONS
Administrator for Monitoring and Technical Support
Assistant Administrator for R&D
Environmental Protection Agency
401 M Street SW
Washington, DC 20460
Telephone: (202) 426-9454

Description: Programs are geared to industrial pollution control. Involved in programs to determine best technology transfer programs and methods for the dissemination of technology information.

DEPARTMENT OF HEALTH AND HUMAN SERVICES

OCCUPATIONAL SAFETY AND HEALTH TECHNOLOGY TRANSFER
NIOSH
Center for Disease Control
Department of Health and Human Services
5600 Fishers Lane
Rockville, MD 20857
Telephone: (301) 443-2140

Description: Develops criteria for occupational safety and health standards. Technology Transfer is implemented through training activities and through publication of the results of research activities. Cooperates with associations and organizations on meetings pertaining to occupational safety and health.

DEPARTMENT OF DEFENSE

RESEARCH AND ADVANCED TECHNOLOGY
Under Secretary of Research and Engineering
Pentagon
Washington, DC 20301
Telephone: (202) 695-5036

Description: DOD laboratories are active in the Federal Laboratory Consortium. Through this working group, technology transfer interacts with other Federal agencies and potential users at the federal, state and local levels. A publication entitled "Directory of Federal Technology Transfer" is available from GPO. Other publications of interest in the area of technology transfer are "Navy Fact Sheet," and "News Items" published by the Federal Laboratory Consortium.

Coal Assistance Programs

DEPARTMENT OF ENERGY

COAL LOAN GUARANTY PROGRAMS

Office of Coal Loan Programs
Industrial and Utility Applications and Operations
Assistant Secretary Resource Applications
Department of Energy
12th and Pennsylvania Avenue NW
Washington, DC 20461
Telephone: (202) 633-9154

Description: The Coal Loan Guarantee program assists in long term financing for small mine operators in low sulfur coal production. Assists in financing for coal preparation plant projects.

ENERGY TECHNOLOGY R&D PROGRAMS

Fossil Energy
Department of Energy
Washington, DC 20545
Telephone: (301) 343-2784

Description: Provides assistance in the form of grants, research and contracts. Assistance may provide for a variety of cost sharing and incentive arrangements including awards to individual inventors, joint ownership of industry and federal cost shared demonstration projects. Support may be provided for conversion of coals to liquids or gases, clean combustion of coals, on site recovery of fossil energy products from coal, electric power generation utilizing fossil fuels, mining research, support technologies in the conversion and utilization of fossil energy resources. Cities, counties, and state governments, private, profit, non-profit organizations, individuals, and universities are eligible.

UNIVERSITY COAL RESEARCH LABORATORY PROGRAM
Office of Advanced Research & Technology
Office of Fossil Energy
Department of Energy
Washington, DC 20545
Telephone: (301) 353-2784

Description: The University Coal Research Laboratory Program involves coal research at universities with existing laboratories capable of coal research. Laboratories are selected from universities and colleges. Research will include technical problems, economic, environmental and health problems.

Laboratories will develop an integrated research and academic training program on the problems of coal relevent to both national and regional concerns. Each laboratory will research expanded coal utilization or long range alternative methods of liquefying coal.

DEPARTMENT OF INTERIOR

GRANTS FOR REGULATION OF SURFACE COAL MINING AND SURFACE EFFECTS OF UNDERGROUND COAL MINING
State and Federal Programs
Office of Surface Mining Reclamation and Enforcement
Department of Interior
18th and C Streets NW
Washington, DC 20240
Telephone: (202) 342-4225

Description: Grants for state government designated agencies, small operators and coal mining operations. Grants benefit state agencies responsible for regulating reclamation and enforcement. Grants for incremental state costs may cover inspections, equipment, and permanent program development. Technical assistance is on any aspect of regulating surface impacts of coal mining.

The Small Operator Assistance Program is available to eligible operators (less than 100,000 tons annual coal production). Limited to providing a determination of probable hydrologic consequences and a statement of the results of test borings and core samples.

DEPARTMENT OF LABOR

MINE HEALTH AND SAFETY GRANTS AND ASSISTANCE

Assistant Secretary for Mine Safety & Health Administration
Department of Labor
Ballston Towers #3
Arlington, VA 22203
Telephone: (703) 235-8264

Description: Provides grants for research and planning studies designed to improve workmen's compensation and occupational disease laws, and assure an adequate and competent staff of trained inspectors. Any mining state is eligible through their mine inspection or safety agency.

Other assistance programs include the following:

> *Mine Health and Safety Counseling and Technical Assistance* — Provides technical advice, special studies, investigations, development of state mine health and safety programs, with enforcement authority. Any state organization or person interested in safe mining conditions may apply.
>
> *Mine Health and Safety Education and Training* — Assistance is used by mine operators who may need guidance in preparation of mine training plans and by miners who are required to be trained in certain equipment procedures, systems or emergency situations. Any mine operator, miner or agent is eligible for the training.

DEPARTMENT OF HEALTH AND HUMAN SERVICES

COAL MINERS RESPIRATORY IMPAIRMENT TREATMENT CLINICS AND SERVICES

Bureau of Community Programs
Health Services Administration
Department of Health and Human Services
5600 Fishers Lane
Rockville, MD 20857
Telephone: (301) 443-5033

Description: Funding for clinics to develop in areas where there are inactive and active miners. The project will serve a significant number of coal workers with pulmonary impairment without regard for their ability to pay.

SMALL BUSINESS ADMINISTRATION

MINE SAFETY AND HEALTH LOANS

Office of Disaster Loans
Small Business Administration
1441 L Street NW
Washington, DC 20416
Telephone: (202) 653-6376

Description: Provides direct loans and guaranteed/insured loans to make additions to or alterations in the equipment, facilities or methods of operations of small mines. This will enable small mines to meet requirements imposed by the Federal Mine Safety and Health Amendments Act. For eligibility there must be 250 or fewer employees and in receipt of notice of deficiency from MSHA.

APPALACHIAN REGIONAL COMMISSION

APPALACHIAN MINE AREA RESTORATION

Appalachian Regional Commission
1666 Connecticut Avenue NW
Washington, DC 20235
Telephone: (202) 673-7874

Description: Provides grants to rehabilitate areas damaged by mining practices, to help abate mine drainage pollution, seal voids in abandoned coal mines, and extinguish underground mine fires. This activity is limited to federal, state or local bodies of government.

Information on Coal Resources, Supplies and Transportation

DEPARTMENT OF ENERGY

ASSISTANT SECRETARY FOR RESOURCE APPLICATIONS
Department of Energy
12th and Pennsylvania Avenue NW
Washington, DC 20461
Telephone: (202) 633-9222

Description: Responsible for coal resources. Offices deal with high Btu coal gas, low/medium Btu gas and industrial atmospheric fluidized bed combustion. Has the responsibility to develop coal supplies and to deal with the commercial applications.

Specific offices that deal with coal supplies:

Office for Coal Resource Management – Deals with high Btu coal gasification, low/medium Btu coal gasification, fluidized bed coal combustion and direct combustion.

Office of Coal Supply Development – Deals with coal loans, licensing, logistics and production and technology. Assesses coal supply and demand, and commercial application of new coal mining technology.

ASSISTANT SECRETARY FOR FOSSIL ENERGY
Department of Energy
1000 Independence Avenue SW
Washington, DC 20585
Telephone: (301) 353-2642

Description: The Assistant Secretary for Fossil Energy coordinates efforts in developing technologies for increasing the supply of domestic fossil fuels (coal, petroleum, natural gas, oil shale) to supplement existing supplies.

ENERGY INFORMATION ADMINISTRATION
Department of Energy
12th and Pennsylvania Avenue NW
Washington, DC 20461
Telephone: (202) 634-5610

Description: Provides information on coal statistics, supplies and studies. Specific offices with coal data:

> *Office of Energy Data and Interpretation* – This office is within the Office of Energy Data Operations. The office collects, interprets and publishes various data pertaining to the U.S. supply and demand for fossil fuels. This includes coal, crude oil, coke, petroleum products, natural gas, oil shale, peat, helium and hydrogen, and fuel production and consumption. It answers inquiries, provides magnetic tape services, distributes some publications and make referrals.
>
> Holdings include computerized data for imports of heavy fuel oils, imports of crude, finished and unfinished petroleum products, fuels and energy consumption by state and consuming sector and U.S. coal reserves by state and coal bed. In addition 22 fuel industry forms are automated.
>
> *Coal and Electric Power Analysis Division* – This Division is within the Office of Energy Source Analysis within the Office of the Assistant Administrator for Applied Analysis. Studies various aspects of coal and electric power including analyses and forecasts of supply/demand balances of coal. Studies explorations of new technologies to derive energy from coal. Analyzes and forecasts electric power supply and consumption. This includes input fuels by electric utility sector, utility pricing and interfuel substitution. Provides detailed examinations of existing and proposed legislation affecting coal and electric power production.

DEPARTMENT OF INTERIOR

BUREAU OF MINES
U.S. Department of Interior
2401 E Street NW
Washington, DC 20241
Telephone: (202) 634-1310

Description: The Bureau collects, compiles, analyzes and publishes statistical and economic information on all phases of mineral resource development. This includes exploration, production, shipments, demand, stocks, prices, imports and exports. Special studies are made on the effects of potential economic, technologic or legal developments on resource availability.

Mineral and Coal-Related Offices:

Mineral Information Systems and Division of Mineral Availability – Located within the Minerals Information and Analysis Directorate. Contains information covering all aspects of reserves, production, processing, consumption, and international trade. Publishes "Mineral Industry Surveys," "Minerals and Materials," "Mineral Trade Notes," "International Coal Trade," "Commodity Data Summaries," "Status of the Mineral Industries," "Minerals in the U.S. Economy," "Mineral Trends and Forecasts," "Mineral Yearbook" and "Mineral Facts and Problems."

Office of Field and Environmental Operations – Concerned with land assessments of wilderness candidate areas, RARE II lands and Indian lands. Information is available on the impact of proposed projects on mineral supplies, availability of minerals, capital and operating costs from identified deposits. Holds the Minerals Availability Systems and produces technical reports and data compilations.

Bureau of Mines Alaska Field Operations Center
P.O. Box 550
Juneau, Alaska 99801
Telephone: (907) 364-2111

Description: Conducts engineering studies of mineral and energy reserves and resources in Alaska. Evaluates and interprets production methods to determine the most effective practices for production in Alaska. Performs investigations designed to promote development of these resources and develops special programs for environmental improvement. Interests include mining, geology, mining costs, mineral deposits and resources, environmental studies, mineral explorations and energy resource evaluations.

Bureau of Mines Eastern Field Operations Center
4800 Forbes Avenue
Pittsburgh, PA 15213
Telephone: (412) 621-4500

Description: Performs engineering and technical investigations to evaluate mineral reserves, mineral land assessments, technological and environmental aspects of mining and the mineral information activities in the 26 states east of the Mississippi River. Holdings include computerized data bases on bituminous coal, mineral, mining operation locations and mine maps.

Bureau of Mines Western Field Operations Center
East 315 Montgomery Avenue
Spokane, WA 99207
Telephone: (509) 484-1610

Description: Collects and analyzes data relating to mineral deposits, mineral industries, mineral economics, deposit evaluation, and mining processing costs primarily in the states of Oregon, Washington, Idaho, Montana, California and Nevada. Holdings include engineering and geological publications of the Bureau of Mines, U.S. Geological Survey and western state bureaus. The center has access to computerized files, including the Mineral Information and Location System (a file of mineral properties and processing plants relating to mineral production) and the Minerals Availability System (a file that covers all mineral resources available to domestic industries.

NATIONAL COAL RESOURCES DATA SYSTEM

Branch of Coal Resources
Geologic Division
U.S. Geological Survey
Department of Interior
National Center
Reston, VA 22092
Telephone: (703) 860-7464

Description: NCRDS has information on the distribution and quality of coal resources. Contains information on pertinent geologic, geochemical, engineering and mining, drill hole, geodetic resources and production data. Data are also provided on petrology and trace elements in coal.

Answers brief general inquiries free. More extensive inquiries are answered as time permits at cost to the requestor. Holdings include computer based coal related data. Permits remote terminal access of coal data at cost to user.

CENTRAL REGION, U.S. GEOLOGICAL SURVEY

Department of Interior
Box 25046
Federal Center
Denver, CO 80225
Telephone: (303) 234-3736

Description: The Central Region is organized into 7 divisions. They assess the nation's mineral and energy resources. The Energy Resources Program has coal geologists evaluating the quantity and quality of coal. Regional studies include the distribution of coal, coal's physical and chemical properties and potential mining hazards. The Mineral Resources Program studies mineral exploration and mineral resource policies. The Geochemistry and Geophysics Program researches uranium and other deposits, geothermal energy, coal and permafrost.

EARTH RESOURCES OBSERVATION SYSTEM
Data Center
U.S. Geological Survey
Department of Interior
Sioux Falls, SD 57198
Telephone: (605) 594-6511

Description: The data center receives information from the Department of Interior, NASA, Department of Defense and other sources to provide training and assistance in the application of remotely sensed data to resources and land use problems. Conducts research in the extraction of information from data and in the interpretation of data for natural resources and land use investigations. The center's areas of interest include energy resources, engineering site selection, geography, geology, land use, mineral resources and natural resources.

DEPARTMENT OF TRANSPORTATION

RESEARCH AND SPECIAL PROGRAMS ADMINISTRATION
Department of Transportation
400 7th Street SW
Washington, DC 20590
Telephone: (202) 426-4461

Description: The purpose of the program is to research and analyze the safety of various materials utilized in transportation systems.

Specific source for information on coal transportation:

> *Transportation Systems Center* – The center assesses the compatibility of the national transportation system with available national resources and advanced systems for the movement of goods.

FEDERAL RAILROAD ADMINISTRATION
Department of Transportation
400 7th Street SW
Washington, DC 20590
Telephone: (202) 426-0881

Description: Conducts studies and research on future requirements for rail transportation. R&D is aimed at providing primary techniques and technology for use by railroads, railroad equipment suppliers, state and local governments. Research involves improved rail freight service, freight car management, safety, policy planning and program evaluation, and development of state rail planning methodology. FRA conducts studies on the economics of railroad transportation. Studies are performed on car shortage problems and rail network capabilities.

DEPARTMENT OF DEFENSE

DEFENSE LOGISTICS AGENCY
Cameron Station
5010 Duke Street
Alexandria, VA 22314
Telephone: (202) 274-4617

Description: Procures coal and other fuels for the Defense Department. Buys fuel burning equipment, liquid propellants, fuel oils, ores, minerals, natural and synthetic solid fuels.

GENERAL SERVICES ADMINISTRATION

FEDERAL SUPPLY SERVICE
General Services Administration
Crystal Mall
1941 Jefferson Davis Highway
Arlington, VA 20406
Telephone: (703) 472-1701

Description: Reviews management of coal resources. Assesses coal needs for government buildings. Studies how government buildings can be fuel efficient and practice conservation.

APPALACHIAN REGIONAL COMMISSION

APPALACHIAN REGIONAL COMMISSION
1666 Connecticut Avenue NW
Washington, DC 20235
Telephone: (202) 673-7869

Description: The Commission is composed of the governors of the 13 states that comprise Appalachia. In November 1977, ARC completed a study of the impact of coal hauling on the region's roads. The report is entitled "An Assessment of the Effects of Coal Movement on the Highways of the Appalachian Region." In January 1978, DOT released a national study of energy roads which concluded that the problem of building or rebuilding roads to haul coal was concentrated in the Appalachian region and that a national program was needed to deal with the total problem.

Sources for Environmental and Safety Information

DEPARTMENT OF ENERGY

ASSISTANT SECRETARY FOR ENVIRONMENT

1000 Independence Avenue SW
Washington, DC 20585
Telephone: (202) 252-4700

Description: Responsible for ensuring that the implementation of all departmental programs is consistent with environmental and safety laws, regulations, and policies. Provides policy guidance to ensure compliance with environmental protection laws and is responsible for reviewing and coordinating all environmental impact statements prepared within DOE. Conducts environmental and health related research and development programs, such as studies of energy-related pollutants and their effects on biological systems. The Office coordinates DOE's responsibilities under the National Environmental Policy Act.

ENVIRONMENTAL ASSESSMENTS AND IMPACT STATEMENTS

Department of Energy
1000 Independence Avenue SW
Washington, DC 20585
Telephone: (202) 252-6020

Description: Copies of DOE Environmental assessments and draft environmental impact statements may be seen in the DOE public reading room and at the following locations: Albuquerque Operations Office, Chicago Operations Office, Idaho Operations Office, Nevada Operations Office, Oak Ridge Operations Office, Richland Operations Office, San Francisco Operations Office, and Savannah River Operations Office. A limited number of single copies are available from

Technical Information Center, Oak Ridge, TN.

In individual cases, copies of environment assessment and draft environmental impact statements may be seen at the following DOE Technology Centers: Bartlesville Energy Technology Center, Grand Forks Energy Technology Center, Laramie Energy Technology Center, Morgantown Energy Technology Center, and Pittsburgh Energy Technology Center.

INFORMATION CENTER FOR ENERGY SAFETY
Oak Ridge National Laboratory
P.O. Box X
Oak Ridge, TN 37830
Telephone: (615) 574-0377

Description: Provides a means for collecting, storing, evaluating and disseminating relevant safety information essential to the development and use of various forms of energy. Publishes state of the art reports and bibliographies.

ECOLOGICAL SCIENCES INFORMATION CENTER
Oak Ridge National Laboratory
P.O. Box X
Oak Ridge, TN 37830
Telephone: (615) 626-2759

Description: Assesses the environmental impact of both nuclear and fossil energy. Publishes a literature overview on the Environmental Health and Control Aspects of Coal Conversion.

Reviews and bibliographies are available from NTIS. In-depth literature searches using computerized data bases and extensive library facilities are provided as is literature support to many environmental research projects related to energy.

HEALTH AND ENVIRONMENTAL STUDIES PROGRAM
Information Center Complex
Oak Ridge National Laboratory
P.O. Box X
Building 7509
Oak Ridge, TN 37803
Telephone: (615) 574-7794

Description: Provides information on toxic substances. Toxic substance information includes production and use, chemical and physical properties, metabolical and physiological, toxicological, environmental and ecological effects.

DEPARTMENT OF INTERIOR

DIVISION OF MINERALS HEALTH AND SAFETY TECHNOLOGY

Minerals Research Directorate
Bureau of Mines
Department of Interior
2401 E Street NW
Washington, DC 20241
Telephone: (202) 634-1210

Description: Areas of interest include R&D in mining technology to protect the occupational health and safety of miners, and to develop safety standards for mines. Publishes technical reports and provides information on R&D in progress.

HEALTH AND SAFETY ANALYSIS CENTER

Department of Interior
P.O. Box 25367
Denver, CO 80225
Telephone: (303) 234-3225

Description: The analysis center provides information on mines, engineering, accidents, explosive gases and fires, statistical data, and disseminates technical and engineering finds. Publishes *Basic Standard Injury and Worktime Statistics on the Mineral Industries Analyses of Factors Related to Mine Accidents.*

NATIONAL MINE HEALTH AND SAFETY ACADEMY

Department of Interior
P.O. Box 1166
Berkley, WV 25801
Telephone: (304) 255-0451

Description: Trains coal and metal/nonmetal mine inspectors and other technical specialists. The Academy offers training in health and safety management.

ENVIRONMENTAL PROTECTION AGENCY

ENVIRONMENTAL PROTECTION AGENCY

401 M Street SW
Washington, DC 20460
Telephone: (202) 755-0700

Description: Coordinates and supports research and antipollution activities by state and local governments, private and public groups, individuals, and educational institutions. The air activities include development of programs, policies, and regulations for air pollution control. EPA is responsible for developing national standards for air quality, emission standards for new stationary sources, and emission standards for hazardous pollutants. EPA is involved in programs in the areas of occupational health and safety, analysis of industrial systems, energy policies and systems, and energy R&D.

Sources for Specific Environmental Information:

Environmental Impact Statements – The statements are available for review. The statements are filed as announced in the *Federal Register*. For a copy of a statement contact the federal agency that prepared the EIS.

Air Pollution Control-Technical Information Services – This is available from the Library Services Office in Research Triangle Park, NC 27711. Literature searches are made for state and local governmental air pollution control agencies, industry, EPA grantees and contractors.

National Ambient Air and Source Emission Data (SAROAD/NEDS) – Available from the National Air Data Branch, of the Monitoring and Data Analysis Division of the Office of Air Quality Planning and Standards at Research Triangle Park, NC 27711. Provides data processing and analysis of raw air quality and source inventory data submitted by state and local air pollution control agencies or collected by EPA. Reports and publications are available.

DEPARTMENT OF LABOR

MINE SAFETY AND HEALTH ADMINISTRATION
Department of Labor
Ballston Tower #3
4015 Wilson Blvd.
Arlington, VA 22203
Telephone: (703) 235-1452

Description: Areas of interest include coal, metal and nonmetal mine health and safety. Responsible for the development and promulgation of mandatory safety and health standards.

Publications include annual reports, informational reports, safety reviews, instruction guides, special reports, and health and safety specifications pertaining to the mineral industries.

Specific Office for Coal Safety:

> *Office of Coal Mine Health and Safety* – Responsible for the administration of the Coal Mine Health and Safety Act of 1969.

OFFICE OF WORKERS COMPENSATION PROGRAMS
Department of Labor
200 Constitution Avenue NW
Washington, DC 20210
Telephone: (202) 523-6692

Description: Responsible for the black lung benefit payment provisions of the Federal Coal Mine Health and Safety Act of 1969. This extends benefits to coal miners and survivors who are totally disabled due to pneumoconiosis, a respiratory disease contracted after prolonged inhalation of coal dust.

DEPARTMENT OF AGRICULTURE

PUBLIC AFFAIRS
Department of Agriculture
14th and Independence Ave. SW
Washington, DC 20250
Telephone: (202) 447-7977

Description: Interested in improving the environment while protecting land and natural resources.

Specific departments that deal with environmental problems pertaining to coal:

> *Office of Energy* – Responsible for developing and coordinating the department's energy policies and programs as it concerns energy and land use.
>
> *Agricultural Research Service* – Researches natural resources, environmental conditions and how to rejuvenate strip mined lands.

DEPARTMENT OF HEALTH AND HUMAN SERVICES

NATIONAL INSTITUTE FOR OCCUPATIONAL SAFETY AND HEALTH
Center for Disease Control
Department of Health and Human Services
Parklawn Building
Rockville, MD 20857
Telephone: (301) 443-1530

Description: NIOSH is a component of the Center for Disease Control. NIOSH administers programs to provide safe and healthy working conditions. Develops safety standards and carries out research in the field.

Specific analysis center for environmental data:

> *Environmental Analysis Center* – The Center is a clearinghouse within NIOSH. Covers all aspects of occupational safety and health with information available on the environment.

NATIONAL INSTITUTE OF ENVIRONMENTAL HEALTH SCIENCES
Department of Health and Human Services
Research Triangle Park, NC 27711
Telephone: (919) 541-3201

Description: Conducts research concerned with defining, measuring, and understanding the effects of chemical, biological, and physical factors in the environment.

DEPARTMENT OF DEFENSE

DEPUTY ASSISTANT SECRETARY FOR ENGINEERING, ENVIRONMENT, AND SAFETY
Assistant Secretary of Defense for Manpower, Reserve Affairs and Logistics
Pentagon
Washington, DC 20301
Telephone: (202) 695-0221

Description: Responsible for policies concerning the safety effects of environmental and energy problems as they relate to DOD programs and facilities.

Specific office that reviews energy policy as it pertains to environmental issues:

> *Directorate for Energy* – Develops monetary policies for the use of coal within installations. Responsible for policies that

involve the use of all the fuels that can best serve respective operations, functions, forces, and operations.

COUNCIL ON ENVIRONMENTAL QUALITY

COUNCIL ON ENVIRONMENTAL QUALITY
722 Jackson Place NW
Washington, DC 20006
Telephone: (202) 395-5700

Description: The Council develops and recommends to the President national policies which further environmental quality. Performs a continuing analysis of changes and trends in the environment. Performs studies, research and analyses relating to the environment. CEQ has oversight responsibility for the National Environmental Policy Act. This requires all federal agencies to include a detailed environmental impact statement in every recommendation or report on legislative proposals and other major federal actions affecting the quality of the environment. All coal related activities that have a significant impact on the environment and that need federal authorization require an environmental impact statement.

FEDERAL MINE SAFETY AND HEALTH REVIEW COMMISSION

FEDERAL MINE SAFETY AND HEALTH REVIEW COMMISSION
1730 K Street NW
Washington, DC 20006
Telephone: (202) 653-5633

Description: Independent commission that deals with contested enforcement activities. Conducts hearings on mine safety and resolves disputes arising under enforcement of the Federal Mine Safety and Health Act of 1977. In fiscal year 1979, mine operators, miners and their representatives submitted nearly 5,000 requests for disposition on which there were about 2,800 decisions by administrative law judges.

Major decisions by the commission:

Magma Copper Co. vs Secretary of Labor, FMSHRC Docket No. DENV 78-533-M, decided Dec. 10, 1979;

Energy Fuels Corp. vs Secretary of Labor, FMSHRC Docket No. DENV 78-410, decided May 1, 1979;

Secretary of Labor vs Cut Slate, Inc., FMSHRC Docket No. WILK 79-13-P, decided July 25, 1979;

Secretary of Labor vs Old Ben Coal Co., FMSHRC Docket No. VINC 79-119, decided Oct. 29, 1979;

Secretary of Labor vs Jim Walter Resources, Inc., and Cowin and Co., FMSHRC Docket No. BARB 77-266-P and BARB 77x465-P, decided Nov. 21, 1979;

Secretary of Labor vs Old Ben Coal Co., FMSHRC Docket No. VINC 74-11, IBOMA 75-52, decided Dec. 12, 1979.

Surface Mining and Mine Leasing Information

DEPARTMENT OF ENERGY

LEASING POLICY DEVELOPMENT
Assistant Secretary for Resource Application
Department of Energy
12th and Pennsylvania Avenue NW
Washington, DC 20461
Telephone: (202) 633-9421

Description: Has jurisdiction and manages the DOE's energy resource programs. Establishes goals for federally owned resources. Studies economic conditions of individual leases. Responsible for leasing policy development.

DEPARTMENT OF INTERIOR

LAND AND WATER RESOURCES
Bureau of Land Management
Department of Interior
18th & C Streets NW
Washington, DC 20240
Telephone: (202) 343-8291

Description: Provides policy guidelines and criteria for management of public lands and resources administered by the Bureau.

Offices responsible for mining policies and resources are as follows:

Office of Coal Management – Responsible for coal operations on public lands, program development and policy analysis within the Bureau.

Division of Mineral Resources – Has jurisdiction over leasing programs for coal, minerals, oil and gas. Involved in minerals management, the minerals data center, mineral resources and mineral environment assessment.

ASSISTANT SECRETARY FOR ENERGY AND MINERALS
Department of Interior
18th and C Streets NW
Washington, DC 20240
Telephone: (202) 343-2186

Description: Responsible for the development of the minerals program, research and development programs and the formulation of the minerals budget.

Offices responsible for surface mining, leasing of public lands are as follows:

Office of Surface Mining Reclamation and Enforcement – Primary goal is to create a program to protect the environment from the adverse effects of coal mining operations while ensuring an adequate supply of coal. Responsible for establishment of minimum standards for regulating the surface effects of coal mining. Administers the abandoned Mine Reclamation Fund that provides for funds to be used for reclaiming and restoring land and water resources adversely affected by past coal mining. Concerned with strip mining regulations.

Specific offices responsible for mining and reclamation:

Office of Abandoned Mine Lands – Administers reclamation of abandoned mine lands. This is accomplished with trust fund monies supported by fees paid for each ton of coal mined.

Office for Technical Services and Research – Oversees federal mine plans and develops improved technology in coal surface mining to comply with environmental protection regulations and administers funds for the Mineral Resource Institute. Provides legal and technical expertise to help resolve problems in the reclamation and coal mining industry.

U.S. Geological Survey – The Branch of Mining Operations within the Conservation Divisions oversees the operations of private industry in mining on the federal lands domain. Oversees leasing, and maintains production accounts and collects royalties.

Sources for International Data

DEPARTMENT OF COMMERCE

NATIONAL TECHNICAL INFORMATION SERVICE

Department of Commerce
5285 Port Royal Road
Springfield, VA 22161
Telephone: (703) 487-4650

Description: Information is available through NTIS to utilize and understand foreign technical data through the following sources:

SFCSIP (Special Foreign Currency Science Information Program) — NTIS coordinates and administers program. SFCSIP produces translated foreign research and development results.

Russian copyrighted scientific articles, books licensed to NTIS — NTIS has negotiated the rights to publish and sell English translations of 6 copyrighted Soviet scientific and technical journals as well as some articles.

Specialty Communications — These communications are from U.S. embassies and foreign service posts to the Department of State. They deal with scientific information being considered for inclusion in the NTIS product line.

Daily Report: People's Republic of China — This report is issued by the Foreign Broadcast Information Service. The information is provided on agriculture, energy and other data.

Engineering Sciences Data Unit — The basic ESDU package is the Data Item. This is a set of looseleaf sheets devoted to a single topic.

Applications of Modern Technology to International Development — Describes new technical reports of particular interest to cooperating agencies in various countries.

Foreign Translations — The Joint Publications Research Service translates and abstracts foreign language, political and technical media for federal agencies. The reports give scientists, engineers, technicians, researchers and business people access to these reports.

Areas of interest covered include People's Republic of China, USSR (Science and Technology), USSR (Economy and Industry), USSR and Eastern Europe, Western Europe, Near East and North Africa, Japan, Latin America and International.

Index to JPRS translations is available from the Micro Photo Division of Bell and Howell, Old Mansfield Road, Wooster, OH 44691.

INTERNATIONAL TRADE ADMINISTRATION
Department of Commerce
14th between E and Constitution Avenue NW
Washington, DC 20230
Telephone: (202) 377-3808

Description: Advises and assists on regulating exports of U.S. goods and technology for the purposes of national security, foreign policy and short supplies. ITA produces international business publications. The series are "Foreign Economic Trends," "Overseas Business Reports," "Global Market Surveys," "Country Market Sectoral Surveys," "International Economic Indicators," "Current Developments in U.S. and Major Foreign Countries," "Market Share Reports" and "Foreign Market Reports."

These publications produce information on mining, mining equipment, energy, energy equipment, engineering, environmental pollution control, and minerals. Country specialists can provide specific market data for coal supplies and mining machinery and equipment.

Publications are available from the Department of Commerce District Offices, NTIS and GPO. GPO has subject bibliographies in the foreign market area that list publications on marketing research, foreign investments, foreign trade and tariffs.

**FOREIGN TRADE DIVISION,
BUREAU OF CENSUS**
Department of Commerce
Washington, DC 20233
Telephone: (301) 763-5140

Description: Collects, tabulates and publishes statistical data on coal exports and imports. The principal products are reports, computer tapes, special tabulations, catalogs and guides to export data. The "Guide to Foreign Trade Statistics" describes sources and types of available data from the Census Bureau.

DEPARTMENT OF ENERGY

**OFFICE OF THE ASSISTANT SECRETARY
FOR INTERNATIONAL AFFAIRS**
Department of Energy
1000 Independence Avenue SW
Washington, DC 20585
Telephone: (202) 252-5800

Description: Directs and manages programs related to the international aspects of overall energy policy. Assists the President with advice on international energy negotiations. The Office assesses world prices, supply trends, and technological developments as they affect the U.S. energy supply. Studies the international movement of coal resources.

DEPARTMENT OF INTERIOR

BRANCH OF FOREIGN DATA
Bureau of Mines
Department of Interior
2401 E Street NW
Washington, DC 20240
Telephone: (202) 632-8970

Description: Provides information in the form of reports on Africa, Middle East, Canada, Australia, Central and South America, China, Japan and Southeast Asia, and Eastern and Western Europe. Data are provided on mining and minerals as they pertain to specific regions.

DEPARTMENT OF STATE

BUREAU OF ECONOMIC AFFAIRS
Department of State
2201 C Street NW
Washington, DC 20520
Telephone: (202) 632-0396

Description: Responsible for recommendations to prevent unlawful business practices that deal with the trading of raw commodities.

Specific bureaus that deal with coal:

> *Bureau of Economic and Business Affairs* – The Fuels and Energy Office within the Bureau is concerned with international energy exports, research, and policies.
>
> *Bureau of Oceans and International Environmental and Scientific Affairs* – Responsible for global policies that deal with environmental health and natural resources. Involved in energy technology, transfer and cooperation.

DEPARTMENT OF TREASURY

OFFICE OF INTERNATIONAL ENERGY POLICY
Assistant Secretary for International Affairs
Department of Treasury
15th and Pennsylvania Avenue
Washington, DC 20220
Telephone: (202) 566-5071

Description: Deals with policies dealing with international energy monetary, commercial, trade policies and programs. Maintains contacts with international organizations.

INTERNATIONAL TRADE COMMISSION

INTERNATIONAL TRADE COMMISSION
701 E Street NW
Washington, DC 20436
Telephone: (202) 523-0146

Description: Researches economic, foreign market data on the trading of fuels. ITC publishes statistical data on coal trade.

Specific office that has information on coal:

> *Office of Industries* – Studies investigations and research projects relating to international trade. This includes minerals, metals, energy and chemicals.

EXPORT IMPORT BANK OF U.S.

EXPORT IMPORT BANK OF U.S.
811 Vermont Avenue NW
Washington, DC 20571
Telephone: (202) 566-8890

Description: Provides credit and financing assistance for U.S. exporters involved in the mining and coal industries. Provides programs such as Export Credit Insurance Program, Cooperative Financing Facility, Commercial Bank Exporter Guarantee Program, Discount Loan Program, and the Export Counseling Services.

ORGANIZATION FOR ECONOMIC COOPERATION AND DEVELOPMENT

OECD PUBLICATIONS AND INFORMATION CENTER
1750 Pennsylvania Avenue NW
Washington, DC 20006
Telephone: (202) 724-1857

Description: Member countries promote economic growth, employment and improved standards of living. Publications are available on international development, environment pollution and waste, energy, industry and science and technology.

UNITED NATIONS

UNITED NATIONS DEVELOPMENT PROGRAM

United Nations
1 United Nations Plaza
New York, NY 10017
Telephone: (212) 754-4790

Description: The program supports projects in the energy and environmental field throughout the world. Researches fuel reserves and energy sources, mining and mineral deposits. The publication "Development Forum Business Edition" provides information on future projects.

ECONOMIC COMMISSION FOR EUROPE

United Nations
Palais des Nations
CH-1211 Geneva 10, Switzerland
Telephone: Geneva (022) 34-60-11

Description: ECE membership consists of all UN member states in Europe, U.S. and Canada. Areas of interest include intergovernmental economic and technical cooperation in transport, trade, steel, coal, electric power, gas, housing building, urban and regional research, agriculture, timber, statistics, water resources, chemicals, engineering, science and technology, environmental problems and long-term perspectives of general energy problems.

Publications include "Economic Survey of Europe," "Economic Bulletins for Europe," and approximately 100 other reports and statistical bulletins. ECE studies are made available through the Sales Section of the U.N. in New York, and Palais des Nations, Geneva.

WORLD BANK

WORLD BANK

1818 H Street NW
Washington, DC 20433
Telephone: (202) 477-1234

Description: Makes loans to governments and private industry in member countries. Involved in issues concerning coal resources in developing countries. World Bank has three publications available entitled "Coal Development Potential and Prospects in the Developing Countries," "Energy Options and Policy Issues in Developing Countries," and "The Mining Industry and the Developing Countries." World Bank also publishes "Operational Summary of Proposed Projects." This summary lists projects forthcoming in the energy and environmental field as long as two years in the future.

INTERNATIONAL ENERGY AGENCY

INTERNATIONAL ENERGY AGENCY COAL RESEARCH

Economic Assessment Service
NCB(IEA) Services Ltd.
14/15 Lower Grosvenor Place
London SW1W OEX, England
Telephone: 01-828-4661

Description: The Economic Assessment Service evaluates potential uses of coal internationally. The purpose is to assess the impact of current and projected utilization technologies and to help member countries of the International Energy Agency. The IEA formulates new projects tailored to the economic, social and physical resources of each member country. Areas of interest include energy economics, impact of coal technologies on coal production, supply, conversion and the economics of pollution control.

Publications include plans and projections for coal production, trade and consumption. Reports are available on the economic and technical criteria for coal utilization plants, costs of control of sulfur pollution, economics of power generation and coal conversion. Publishes *Coal Abstracts.*

Evaluates and analyzes data free for participating countries. May distribute publications at a fee to other approved institutions.

EUROPEAN COMMUNITIES COMMISSION

EUROPEAN COMMUNITIES COMMISSION

European Communities Information Service
2100 M Street
Suite 707
Washington, DC 20037
Telephone: (202) 862-9500

Description: Areas of interest include the activities and policies of the European community. This includes European Economic Community, European Coal and Steel Community and the European Atomic Energy Community. Publishes bulletins and brochures on the various activities. A list of publications is available.

NEW SOUTH WALES DEPARTMENT OF MINES

NEW SOUTH WALES DEPARTMENT OF MINES
Geological Survey of New South Wales
State Office Block
Phillip Street
Sydney, N.S.W.
2000 Australia

Description: Responsible for mapping and assessment of the geology and mineral resources of New South Wales. Conducts geological research and maintains the State Core Library.

Publications include "Records of Geological Survey of N.S.W.," "Bulletins," "Technical Reports," "Mineral Resources and Mineral Industry Series," and "Mine Data Sheets."

Holdings include special collection of geological literature. State Core Library contains samples and cores from exploratory boreholes in New South Wales. The Survey is a repository for unpublished reports from mineral exploration companies in N.S.W. Indexes to published and unpublished geological reports and a Core Library index are available for consultation by the public.

GEOLOGICAL SURVEY OF CANADA

DEPARTMENT OF ENERGY, MINES AND RESOURCES
Institute of Sedimentary and Petroleum Geology
Geological Information Library
Geological Survey of Canada
3303 33rd Street NW
Calgary, Alberta, Canada T2L2AL
Telephone: (403) 284-0301

Description: Areas of interest include geology, geophysics, soils, geochemistry, petroleum, coal, natural gas, mineral resources, paleontology, palynology, micropaleontology, and energy.

Answers inquiries and makes referrals to other sources of information. Services are free except for reproduction.

NATIONAL COUNCIL FOR U.S. CHINA TRADE

NATIONAL COUNCIL FOR U.S. CHINA TRADE
1050 17th Street NW
Washington, DC 20036
Telephone: (202) 828-8300

Description: Nonprofit private association that maintains information for its members. Its purpose is to facilitate and promote trade between the People's Republic of China and the U.S. Offers import and export advisory services. A committee was formed to research and advise on minimg machinery and equipment.

Publishes special reports, market surveys, directories, "China Business Review," and other publications.

University Sources

COLORADO SCHOOL OF MINES

COLORADO SCHOOL OF MINES
Golden, CO 80401
Telephone: (303) 279-0300

Description: Studies and performs research on mining and the environmental effects from mining. Publications include data on current developments in geology, mining and the minerals industry.

DARTMOUTH COLLEGE

SYSTEM DYNAMICS GROUP
Thayer College of Engineering
Dartmouth College
Hanover, NH 03755
Telephone: (603) 646-3551

Description: The group is sponsored by the federal and state governments, corporations and foundations. Areas of interest include development of computer simulation models that integrate information on social, economic, environmental, geological and political factors governing long-term resource availability. Data are available on coal use and the U.S., forestry management and environmental resource assessment.

Publishes books, reports, articles, abstracts, indexes, bibliographies and reprints. A publications list is available. Answers inquiries, provides information on research in progress and distributes publications. Services are free except for publications.

GEORGIA INSTITUTE OF TECHNOLOGY

ENGINEERING EXPERIMENT STATION

Office of International Programs
Georgia Institute of Technology
Atlanta, GA 30332
Telephone: (404) 894-3875

Description: Areas of interest include small industry development, appropriate technology, alternate energy sources, coal technology transfer and industrial extension. Publications include "International Informer," "Small Development Network," annual and final project reports, reports of research findings, and case histories.

Answers reasonable inquiries free. Provides duplication and consulting services for a fee.

IOWA STATE UNIVERSITY

RARE EARTH INFORMATION CENTER

Energy and Mineral Resources Research Institute
Iowa State University
Ames, IA 50011
Telephone: (515) 294-2272

Description: The Center is sponsored by the Department of Energy, foreign rare earth producers and advanced technology corporations. The Center is concerned primarily with the physical metallurgy and solid state physics of the rare earth metals and their alloys. It maintains files on the analytical, inorganic and physical chemistry, geochemistry, and toxicity of the rare earth elements and compounds. Publications include *Rare Earth Information Center News.*

MASSACHUSETTS INSTITUTE OF TECHNOLOGY

SCHOOL OF ENGINEERING
M.I.T.
77 Massachusetts Avenue
Cambridge, MA 02142
Telephone: (617) 253-3418

Description: Compiled the internationally sponsored study entitled "World Coal Study." The coal producing countries along with M.I.T. produced the report and stated that goals can be reached through a 5% annual growth in coal production. Improvements in transportation and equipment are needed. Government activities to stabilize environmental standards and encourage investments are also needed.

PENNSYLVANIA STATE UNIVERSITY

COAL RESEARCH SECTION
College of Earth and Mineral Sciences
Pennsylvania State University
University Park, PA 16802
Telephone: (814) 865-6545

Description: The Coal Research Section is partially sponsored by the U.S. Department of Energy, NSF and the Pennsylvania Science and Engineering Foundation. Areas of interest include characterization of U.S. coals (carbon, combustion and high-temperature gas reactions).

Publications include research summaries. Holdings include numerical data base on in-depth characterization of U.S. coals. Available bibliographic coal data base has 40,000 articles in the carbon coal area. Answers inquiries, conducts seminars and workshops.

DEPARTMENT OF GEOSCIENCES
College of Earth and Mineral Sciences
Pennsylvania State University
University Park, PA 16802
Telephone: (814) 865-6711

Description: Areas of interest include geochemistry, geology, geophysics, mineralogy, coal, minerals, geological surveys.

Publishes technical reports, journal articles, and R&D summaries. Answers inquiries, provides advisory services and information on research in progress. Conducts seminars, analyzes data and makes referrals.

MINE DRAINAGE RESEARCH SECTION
College of Earth and Mineral Sciences
121 Mineral Science Bldg.
University Park, PA 16802
Telephone: (814) 863-1641

Description: This section was created to encourage and coordinate university research activities concerned with coal mine drainage as an environmental problem and a major production consideration of the coal industry. Areas of interest include coal mine acids, neutralization, water treatment, pollution prevention, processes, economics, water quality, sludge disposal, coal mining, surface mining and coal preparation.

Publications include research reports. Holdings include collection of specialized publications concerned with coal mine drainage. Not available for distribution or public use.

Answers inquiries, makes referrals, and provides advisory services. Provides information on research in progress, evaluates data, distributes data compilation and conducts conferences.

PURDUE UNIVERSITY

UNDERGROUND EXCAVATION AND ROCK PROPERTIES INFORMATION CENTER
Center for Information and Numerical Data Analysis and Synthesis
Purdue University
Purdue Industrial Research Park
2595 Yeager Rd.
West Lafayette, IN 47906
Telephone: (317) 463-1581 or
(800) 428-7675

Description: The Center is sponsored by NSF and operated by the Thermophysical Properties Research Center at Purdue University. The Center catalogues and compiles data on mechanical, thermal, and electrical properties of rock formations and substances. Studies the responses of earth medial and rock types to nuclear blast and explosive phenomena. Studies and evaluates excavation and tunneling techniques. The Center has a computer-based data bank of rock properties.

TEXAS A & M UNIVERSITY

CENTER FOR ENGINEERING AND MINERAL RESOURCES

Texas A & M University
College Station, TX 77843
Telephone: (713) 845-8025

Description: Studies information on coal and mineral resources. Coordinates information through publications on Texas mineral and coal resources.

UNIVERSITY OF ALABAMA

STATE MINE EXPERIMENT STATION

Mineral Resources Institute
University of Alabama
University, AL 35486
Telephone: (205) 348-6577

Description: Areas of interest include mineral resources, mining methods, coal technology, mineral preparation, petroleum and water resources, water pollution from minerals and fuels, and drilling and rock mechanics.

Publications include bulletins, technical reports and articles contributed to scientific journals. Answers inquiries and provides consulting services.

UNIVERSITY OF KENTUCKY

ENGINEERING LIBRARY

University of Kentucky
355 Anderson Hall
Lexington, KY 40506
Telephone: (606) 258-2965

Description: Areas of interest include coal research, engineering, engineering mechanics, materials science, environmental science and transportation.

Publishes "Guide to the Engineering Library" and provides reference services. Holdings include special collections on coal research and environmental technology. The library has access to the SDC, Lockheed and SOLINET computerized data bases.

INSTITUTE FOR MINING AND MINERAL RESEARCH
University of Kentucky
P.O. Box 13015
Lexington, KY 40583
Telephone: (606) 252-5535

Description: IMMR conducts a coal R&D program for the Kentucky Department of Energy and provides technical assistance for demonstration projects. It also funds various coal-related projects throughout Kentucky. Areas of interest include energy and natural resources, gasification and liquefaction of coal, coal utilization, reclamation, environmental effects and reserves.

Publishes annual reports and reports on research findings. IMMR has access to all major computerized data bases. Answers inquiries, provides consulting, chemical analysis, literature searching, and conducts economic studies.

UNIVERSITY OF NORTH DAKOTA

UNIVERSITY OF NORTH DAKOTA
University Station
Grand Forks, ND 58202
Telephone: (701) 777-3132

Description: Researches coal resources and the impacts of coal on the environment, coal conversion, gasification and liquefaction.

UNIVERSITY OF WISCONSIN-MADISON

ENVIRONMENTAL MONITORING AND DATA ACQUISITION GROUP
Institute for Environmental Studies
University of Wisconsin-Madison
1063 WARF
610 Walnut Street
Madison, WI 53706

Description: Partially sponsored by the federal government, state government, and private industry. The group conducts research in environmental monitoring. Areas of interest include environmental impact of coal-fired power plants, remote sensing applied to water resources, development of new concepts for cadastral systems, applied research directed at state agency problems, information systems, aerial photography, computer analysis and thermal scanning.

Publications include technical reports, journal articles and special reports. Answers inquiries, makes referrals, provides advisory and magnetic tape services, provides information on research in progress, distributes data compilations and publications.

UTAH STATE UNIVERSITY

ECONOMICS RESEARCH INSTITUTE

Utah State University
Eccles Business Building
Logan, UT 84322
Telephone: (801) 752-4100

Description: Areas of interest include natural resources planning, use and conservation. Studies are done on econometrics, manpower, finance and economics of public land policy.

Publications include working paper series and reprint series. Answers inquiries, makes referrals, and provides consulting services. Research is conducted on a grant basis.

WEST VIRGINIA UNIVERSITY

COAL RESEARCH BUREAU

College of Mineral and Energy Resources
West Virginia University
Morgantown, WV 26506
Telephone: (304) 293-4207

Description: Areas of interest include information on increasing uses and markets for West Virginia. Researches information on coal carbonization, desulfurization, utilization and beneficiation of coal-associated minerals, production of structural products from coal ash and refuse, sewage treatment using coal, acid mine drainage, air pollution, coal preparation and computer applications to mining and coal utilization. The Bureau has complete pilot-scale coal preparation facilities and complete analytical facilities.

Publishes report series issued periodically on Bureau findings. Holdings include publications relating to the coal industry, energy research and waste utilization. The Bureau has a computational center housing and IBM 360/75 computer and CatComp plotting equipment Answers brief inquiries and conducts contract research in areas of interest.

WEST VIRGINIA COAL MINING INSTITUTE

West Virginia University
Morgantown, WV 26506
Telephone: (304) 293-5695

Description: Areas of interest include coal mining, engineering and economics connected with coal mine development and operation. Concerned with mine ventilation and safety.

Publications include transactions of the Institute. Provides copies of the Institute's transactions to members and limited number of libraries without charge and to nonmembers for a fee.

Principal State Coal Organizations

COUNCIL OF STATE GOVERNMENTS

COUNCIL OF STATE GOVERNMENTS
Iron Works Pike
P.O. Box 11910
Lexington, KY 40578
Telephone: (606) 252-2291

Description: The Council is a joint agency of all the state governments. Conducts research on state programs, problems, and maintains an information service available to state agencies, officials and legislators. Assists in state-federal liaison, promotes regional and state-local cooperation and provides staff for affiliated organizations.

Specific publications that provide information on coal:

Integrated Use of Landsat Data for State Resource Management
State Information Needs for Resource Management
Environmental Resource Data: Intergovernmental Management Dimensions
State Issues in Rail Transportation of Coal
Highway Transportation of Coal: The Kentucky Experience
State Taxation of Railroads and Tax Relief Programs
Inter-Regional Seminar on Power and Energy Needs and Environmental Impacts
State Initiatives for Electric Utility Rate Reform
State Responses to the Energy Crisis
Surface Mine Management in the West

ALABAMA

OFFICE OF THE GOVERNOR

State Capitol Building
Montgomery, AL 36130
Telephone: (205) 832-3511

Description: Governor's office has a department for Conservation and Natural Resources.

ALABAMA DEPARTMENT OF INDUSTRIAL RELATIONS

1816 8th Avenue North
Birmingham, AL 35202
Telephone: (205) 251-1181

Description: The division administers state laws concerning coal mine health and safety. Inspects working safety and health conditions in place of employment. Conducts safety, electrical and first aid courses for coal mine personnel. Areas of interest include inspection, investigation, first aid, ventilation, basic electricity, geology, mechanical guarding, gases, applicable equipment, and mine rescue and transportation. Publishes coal report and data compilations.

ALASKA

OFFICE OF THE GOVERNOR

State Capitol, Pouch A
Juneau, AK 99811
Telephone: (907) 465-3500

Description: The governor's office has a number of departments and divisions that are concerned with the changes that will occur through the development of energy sources. These departments and divisions include, the Division of Energy and Power Development, Department of Environmental Conservation, Department of Natural Resources and Planning, Division of Geological and Geophysical Surveys, Division of Land and Water Management and Division of Mineral Management.

ARKANSAS

ARKANSAS DEPARTMENT OF LABOR

Mine Inspection Division
247 Central Mall
Fort Smith, AR 72903
Telephone: (501) 452-5132

Description: Areas of interest include coal mining, mining safety, mining enforcement and mine safety training. Publishes annual report. Conducts mine safety courses for miners.

CALIFORNIA

OFFICE OF THE GOVERNOR
State Capitol
Sacramento, CA 95814
Telephone: (916) 445-2841

Description: The state is interested in the areas of energy needs for the future, emergency plans to deal with shortages, environmental concerns, and all forms of energy that will prove economical for the needs of the state. The state of California has an energy commission that studies conservation, engineering and demand and supply assessment.

COLORADO

OFFICE OF THE GOVERNOR
136 State Capitol Building
Denver, CO 80203
Telephone: (303) 839-2471

Description: The state government has several offices and divisions that deal with energy and natural resources. The Office of Energy Conservation deals with policy and planning. The state has a Division of Mineral and Energy Impacts. The Department of Natural Resources studies policy, planning and the efficient use of land.

IDAHO

OFFICE OF THE GOVERNOR
State Capitol
Boise, ID 83720
Telephone: (208) 334-2100

Description: The Office of Energy within the state government studies conservation resource programs and the utilization of all energy resources within the state. Interest in all energy matters pertaining to national programs as they relate to state energy programs. Although Idaho produces no commercial quantity of coal, geologists have demonstrated a statewide coal reserve base of 4.4 million short tons, ranking Idaho 31st among all states in coal reserves.

ILLINOIS

OFFICE OF THE GOVERNOR
Room 207, Capitol Building
Springfield, IL 62706
Telephone: (217) 782-6830

Description: The state is concerned with environmental problems, natural resources and mining.

ILLINOIS DEPARTMENT OF MINES AND MINERALS
704 William G Stratton Building
Springfield, IL 62706
Telephone: (217) 782-6791

Description: Areas of interest include coal mine safety, oil and gas, metal mining, explosives and land reclamation. Publications include data on mining, regulation, resources, production reports and surface mining rules and regulations.

ILLINOIS DEPARTMENT OF BUSINESS AND ECONOMIC DEVELOPMENT
Division of Energy
222 South College Street
Springfield, IL 62706
Telephone: (217) 782-1926

Description: Areas of interest include coal development, energy data research and analysis, energy conservation, alternate energy development, petroleum products, monitoring and allocation, waste recovery and solar energy. Publishes technical reports and reviews.

ILLINOIS GEOLOGICAL SURVEY
Natural Resources Building
Urbana, IL 61801
Telephone: (217) 344-1481

Description: Interests include information on geology, engineering, environmental geology, mining resources and mining chemistry, geology occurrence, and the resource status and the economic aspects of coal, oil and gas.

INDIANA

OFFICE OF THE GOVERNOR

State House
Indianapolis, IN 46204
Telephone: (317) 232-4567

Description: The state office has a Department of Natural Resources and an Energy Group within the Governor's office.

INDIANA GEOLOGICAL SURVEY

611 North Walnut Grove
Bloomington, IN 47401
Telephone: (812) 337-2862

Description: Areas of interest include geophysics, geochemistry, fossils, industrial minerals, petroleum and coal. Publishes a bibliography and directory.

IOWA

OFFICE OF THE GOVERNOR

State Capitol Building
Des Moines, IA 50319
Telephone: (515) 281-5211

Description: The state government has several offices and departments involved in energy planning for the state. These include the Energy Policy Council, Department of Environmental Quality and the Natural Resources Council.

IOWA GEOLOGICAL SURVEY

Geological Survey Building
123 North Capitol Street
Iowa City, IA 52242
Telephone: (515) 338-1173

Description: Areas of interest include geologic features, coal, mineral resources and geophysical surveys. Publications include technical reports.

IOWA DEPARTMENT OF SOIL CONSERVATION

Division of Mines and Minerals
Wallace State Office Building
Des Moines, IA 50519
Telephone: (515) 281-5774

Description: Information is available on surface mining, mining law, excavations, mineral deposits and coal.

KANSAS

OFFICE OF THE GOVERNOR

Capitol, 2nd Floor
Topeka, KS 66612
Telephone: (913) 296-3232

Description: The state office has an Energy Office and the Kansas Senate has a committee that studies energy and natural resources.

KANSAS GEOLOGICAL SURVEY

University of Kansas
1930 Avenue A, Campus West
Lawrence, KS 66044
Telephone: (913) 864-3965

Description: Areas of interest include mineral resources, environmental geology, operations, research, subsurface geology, geochemical research, mineral product development, and energy analysis. Publications include information on chemicals, energy resources, geology, mineral resources and subsurface geology.

KANSAS MINED LAND CONSERVATION AND RECLAMATION BOARD

c/o State Corporation Commission
State Office Building
Topeka, KS 66612
Telephone: (913) 296-3355

Description: Areas of interest include information on the reclamation of land used for the surface mining of coal in Kansas. Publishes administrative rules and regulations.

KENTUCKY

OFFICE OF THE GOVERNOR

State Capitol
Frankfort, KY 40601
Telephone: (502) 564-2611

Description: The state offices have departments and bureaus involved in energy. These include the Office of Energy, Bureau of Energy Management, Bureau of Energy Research. These offices study conservation, shortages and technology assessment. The Department of Natural Resources and Environmental Protection contains the Bureau of Environmental Protection, Bureau of Natural Resources and the Bureau of Surface Mining Reclamation and Enforcement.

KENTUCKY DEPARTMENT OF MINES AND MINERALS
Coal and Clay Mining Division
P.O. Box 680
Lexington, KY 40501
Telephone: (606) 254-0367

Description: Areas of interest include name, production, type and location of coal and clay mines. Provides data on the safe and correct methods of mining. Publishes a safety bulletin and annual report.

MARYLAND

OFFICE OF THE GOVERNOR
State House
Annapolis, MD 21404
Telephone: (301) 269-3901

Description: The state offices have a Department of Natural Resources that studies environmental problems and is involved in geological surveys.

MARYLAND BUREAU OF MINES
City Building
Westernport, MD 21562
Telephone: (301) 359-3057

Description: The Bureau evaluates coal mining permit applications. Issues and enforces coal mining permits. Publishes regulations, standards, specifications and an annual report.

MISSISSIPPI

OFFICE OF THE GOVERNOR
P.O. Box 139
Jackson, MS 39205
Telephone: (601) 354-7790

Description: The state offices contain the Air and Water Pollution Control Commission and the Fuel, Energy and Management Commission.

MISSISSIPPI GEOLOGICAL, ECONOMIC AND TOPOGRAPHICAL SURVEY
P.O. Box 4915
2525 North West Street
Jackson, MS 39216
Telephone: (601) 354-6228

Description: Areas of interest include economic and academic information on geology, and mineral resources. Publications include bulletins, information series, environmental geology series and journal articles.

MISSOURI

OFFICE OF THE GOVERNOR

Executive Office
Jefferson City, MO 65101
Telephone: (314) 751-3222

Description: The state office has a Missouri Energy Program that deals with fuel technology, program development and fuel allocation.

MISSOURI DEPARTMENT OF NATURAL RESOURCES, DIVISION OF POLICY DEVELOPMENT

1014 Madison Street
Jefferson City, MO 65101
Telephone: (314) 751-4000

Description: Responsible for management of the federal state set aside program for emergency assistance. Basic statistical information is available on energy use by quantity, fuel type, energy conservation, and energy resource development in Missouri. Publishes technical reports and R&D summaries.

MISSOURI DEPARTMENT OF NATURAL RESOURCES, DIVISION OF GEOLOGY AND LAND SURVEY

P.O. Box 250
Rolla, MO 65401
Telephone: (314) 364-1752

Description: Areas of interest include all aspects of geology with special emphasis on geologic mapping, structural geology, metallic and nonmetallic mineral resources, coal, oil and gas, engineering geology and geochemistry. Publishes reports and information series.

MONTANA

OFFICE OF THE GOVERNOR

State Capitol
Helena, MT 59601
Telephone: (406) 449-3111

Description: Concerned with fuel development for energy needs within the state. Involved in R&D activities as they pinpoint the priorities and needs of the state. The state office has a Department of Health and Environmental Sciences and a Department of Natural Resources.

NEBRASKA

OFFICE OF THE GOVERNOR

State Capitol Building
Lincoln, NE 68509
Telephone: (402) 471-2244

Description: The Nebraska Energy Office is involved in conservation, utilization of natural resources and environmental control. Information is provided on general coal utilization, coal transportation as it pertains to state needs and geological surveys.

NEW MEXICO

OFFICE OF THE GOVERNOR

State Capitol
Santa Fe, NM 87503
Telephone: (505) 827-2221

Description: The state offices are involved in energy management, conservation, fuel forecasting, R&D and environmental conservation. The state has a Department of Energy and Minerals that is involved in surface mining, resource development, mine inspections, compiling and assembling data on the coal industry to state, federal and local officials. The Natural Resources Department coordinates the development of a coal resource data base.

NEW MEXICO BUREAU OF MINES AND MINERAL RESOURCES

Campus Station
Socorro, NM 87801
Telephone: (505) 835-5410

Description: Conducts field studies for coal resource characterization and other coal research activities. Publications include geologic maps and resource maps.

NEW YORK

OFFICE OF THE GOVERNOR
State Capitol
Albany, NY 12224
Telephone: (518) 474-8390

Description: State interests include conservation and emergency plans. Works with all the governors in the Northeast to promote joint energy policies. The state legislature has energy committees and committees involved in environmental concerns.

NORTH DAKOTA

OFFICE OF THE GOVERNOR
Capitol Building
Bismarck, ND 58505
Telephone: (701) 224-2204

Description: Concerned with energy conservation and management as it pertains to state policies and needs. The state office has a Coal Impact Office that administers funds from the coal severance tax to impacted areas. The Reclamation Division is responsible for permitting coal mine operations including mining and reclamation plans. The Geological Survey is responsible for exploration permits and geological data on coal.

OHIO

OFFICE OF THE GOVERNOR
State House
Columbus, OH 43215
Telephone: (614) 466-3526

Description: The Ohio Department of Energy is the major data source on energy consumption, utilization and production within the state. Performs data acquisition and processing of the state's energy resources. The Energy Resource Analysis Group analyzes Ohio's coal energy potential and quality. The Ohio Geological Survey has field data and specific information on coal data and resources. Publications include *Coal Production in Ohio 1800-1974* and *Analyses of Ohio Coal.*

OKLAHOMA

OFFICE OF THE GOVERNOR
State Capitol
Oklahoma City, OK 73105
Telephone: (405) 521-2342

Description: Interested in matters concerned with environmental health, conservation and mine safety. The state has a Department of Energy.

MINE INSPECTOR FOR OKLAHOMA
4040 N Lincoln
Oklahoma City, OK 73105
Telephone: (405) 521-3859

Description: The Chief Inspector for Mines is interested in regulations in connection with modern methods of mining affecting the health and safety of miners. Involved in regulations applying to ventilation and other mining hazards. Conducts training schools.

PENNSYLVANIA

OFFICE OF THE GOVERNOR
Main Capitol Building
Harrisburg, PA 17120
Telephone: (717) 787-2500

Description: The state has a Governor's Energy Council. The Council is partially sponsored by the U.S. Department of Energy. The Council provides data input for policy development in such areas as conservation, facility siting and transportation. The state has a Department of Environmental Resources and the Pennsylvania House has a Mines and Energy Management Committee.

SOUTH DAKOTA

OFFICE OF THE GOVERNOR
State Capitol
Pierre, SD 57501
Telephone: (605) 773-3212

Description: The state offices are involved in the status of energy policies, management programs, conservation, fiscal programs and the demand on state energy supplies. The state has an office of Energy Policy. Both the Senate and House have a committee that studies natural resources.

TENNESSEE

OFFICE OF THE GOVERNOR
State Capitol Building
Nashville, TN 37219
Telephone: (615) 741-2001

Description: Studies information on the state's energy resources, needs and consumption. Assesses environmental and conservation programs. Distributes information on energy forecasts and trends. The state has a Department of Conservation with a division for geology and surface mining.

TEXAS

OFFICE OF THE GOVERNOR
Capitol Building
Austin, TX 78711
Telephone: (512) 475-4101

Description: Reviews energy policies, technologies, conservation and fuel supplies. Interested in all the environmental consequences involved in the utilization of all types of fuels. The state government has an Energy and Natural Resources Advisory Council. This council appraises new technology assessments and policies.

UTAH

OFFICE OF THE GOVERNOR
210 State Capitol
Salt Lake City, UT 84114
Telephone: (801) 533-5231

Description: The state studies natural resources along with oil, gas and coal mining.

UTAH GEOLOGICAL AND MINERAL SURVEY

606 Black Hawk Way
Salt Lake City, UT 84108
Telephone: (801) 581-6831

Description: Provides information on energy resources with research data on coal, oil, natural gas, oil impregnated sandstones, oil shale, and geothermal energy sources. Information is provided on mineral resources. This includes mining districts, alteration areas, statistical studies and data on mineral extraction from brines of Great Salt Lake. Data are available on engineering and environmental geology.

VIRGINIA

OFFICE OF THE GOVERNOR

Capitol Building
Richmond, VA 23219
Telephone: (804) 786-2211

Description: Concerned with emergency plans for resource shortages. Studies information on all new forms of energy. Involved in environmental concerns as they pertain to the utilization of natural resources.

VIRGINIA DEPARTMENT OF CONSERVATION AND ECONOMIC DEVELOPMENT

Division of Mined Land Reclamation
Drawer V
Big Stone Gap, VA 24219
Telephone: (703) 523-2925

Description: The division is the regulatory agency for enforcement of surface coal mining (contour stripping) and minerals. Publications include standards, specifications, statistical data regarding mining permits, land disturbance and reclamation of acreage and revegetation of surface areas.

WASHINGTON

OFFICE OF THE GOVERNOR

Legislative Building
Olympia, WA 98504
Telephone: (206) 753-6780

Description: Studies plans to deal with energy supplies and new methods for conservation and environmental control. Produces energy economic data. The state has a Department of Natural Resources that deals with geology and earth management.

WEST VIRGINIA

OFFICE OF THE GOVERNOR

West Virginia Capitol Building
Charleston, WV 25305
Telephone: (304) 348-2000

Description: The state offices are interested in air pollution control, mining, natural resources, industrial use of energy and new methods for energy development. The state has an Air Pollution Control Commission, Fuel and Energy Office, Department of Mines and Department of Natural Resources.

WEST VIRGINIA GEOLOGICAL AND ECONOMIC SURVEY COMMISSION

P.O. Box 879
Mont Chateau Research Center
Morgantown, WV 26505
Telephone: (304) 292-6331

Description: Areas of interest include geology and natural resources. Information deals with environmental geology, geologic hazards and land use, planning, coal, natural gas, petroleum, minerals, and fossils. Publishes a newsletter on current geological research in West Virginia.

WISCONSIN

OFFICE OF THE GOVERNOR

State Capitol
East Madison, WI 53702
Telephone: (608) 266-1212

Description: Administers emergency plans for fuel shortages. Researches and studies information pertaining to policies and economic problems. The state government has a Department of Natural Resources which administers environmental standards and resource management.

WYOMING

OFFICE OF THE GOVERNOR

State Capitol
Cheyenne, WY 82002
Telephone: (307) 777-7434

Description: Studies future energy needs and projects. Distributes data on coal and mineral industries. The state government has a Department for Environmental Quality and a State Geological Survey. The Geological Survey studies the coal and industrial needs for the state. The Department of Economic Planning and Development promotes the development of the mineral industry within the state. Provides production data and mineral projections by county. The state Inspector of Mines compiles production and manpower statistics, and safety and accident statistics.

Associations with Coal-Related Information

AMERICAN COKE AND COAL CHEMICALS INSTITUTE
300 N Lee Street
Alexandria, VA 22314
Telephone: (703) 548-8250

Description: Areas of interest include coal, coke, coal chemicals, coke-oven gas, tar, ammonia, naphthalene, benzol, xylol and fuel. Publishes "Foundry Facts" and "Making Efficient Use of Coke in the Cupola."

AMERICAN GEOLOGICAL INSTITUTE
5205 Leesburg Pike
Falls Church, VA 22041
Telephone: (703) 379-2480

Description: Researches information pertaining to geological specialties. Provides publications and information on coal mining and earth sciences.

AMERICAN INSTITUTE OF MINING AND METALLURGICAL AND PETROLEUM ENGINEERS
345 E 47th Street
New York, NY 10017
Telephone: (212) 644-7680

Description: Areas of interest include coal mining and coal mine engineering.

AMERICAN MINING CONGRESS
1200 18th Street NW
Washington, DC 20036
Telephone: (202) 861-2800

Description: Areas of interest include coal mining and all coal mining operations as pertaining to the industry and its policies.

BITUMINOUS COAL OPERATORS ASSOCIATION
303 World Center Building
918 16th Street NW
Washington, DC 20006
Telephone: (202) 783-3195

Description: Areas of interest include labor relations in the bituminous coal industry, mining safety, mining methods, bituminous coal industry's economic conditions and selected legislation. No publications or information circulars are available for distribution. Answers special purpose inquiries and makes referrals.

BITUMINOUS COAL RESEARCH INC.
350 Hochberg Road
Monroeville, PA 15146
Telephone: (412) 327-1600

Description: Affiliate of the National Coal Association. Areas of interest include coal research, mining preparation, combustion, conversion, reclamation of coal mined land, waste material utilization, air and water pollution control, analytical methods, energy resources and use, and fuel engineering.

Publications include "Coal and the Environment Abstract Series," and irregular series. They contain bibliographies and abstracts of holding in mine drainage and reclamation of coal mined lands. Holdings include the Coal Mine Drainage Collection, Reclamation Collection and the Coal Refuse Collection.

COAL EXPORTERS ASSOCIATION
1130 17th Street NW
Washington, DC 20036
Telephone: (202) 628-4322

Description: Areas of interest include coal, coke and coal exports and imports. Publishes monthly bulletin and distributes publications at cost to nonmembers.

EDISON ELECTRIC INSTITUTE
1111 19th Street
Washington, DC 20036
Telephone: (202) 828-7400

Description: Provides information on the coal industry in the form of publications and brochures. Represents electric utility companies.

MINE INSPECTORS INSTITUTE
1900 Grant Building
Pittsburgh, PA 15219
Telephone: (412) 281-2620

Description: Studies mining data of interest to individuals and inspectors involved in mine safety and other problems in the mines.

MINE AND METALLURGY SOCIETY OF AMERICA
230 Park Avenue
New York, NY 10017
Telephone: (212) 687-2752

Description: Interested in the safety aspects of mining and how to best utilize mineral resources.

NATIONAL COAL ASSOCIATION
1130 17th Street NW
Washington, DC 20036
Telephone: (202) 628-4322

Description: Provides information pertaining to the problems of the coal industry. Conducts studies on markets and forecasts and is involved in all aspects of air pollution regulations. Members include coal producers, suppliers, sellers, and transporters.

Publishes "Coal Data," "Coal News," "Coal Traffic Annual," "Steam Electric Plant Factors," "International Coal Members," and "Implications of Investment in the Coal Industry from Other Energy Industries."

NATIONAL COUNCIL OF COAL LESSORS
1130 17th Street NW
Washington, DC 20036
Telephone: (202) 628-4322

Description: Members are owners concerned with the problems of lands with coal deposits.

NATIONAL INDEPENDENT COAL OPERATORS ASSOCIATION
P.O. Box 354
Richlands, VA 24641
Telephone: (703) 963-9011

Description: Members are generally owners of small coal mines. They are involved in safety, employment and mining techniques. Publishes monthly magazine.

SLURRY TRANSPORT ASSOCIATION
490 L'Enfant Plaza SW
Washington, DC 20024
Telephone: (202) 555-4700

Description: Members are involved in transportation of coal. Keeps abreast of developments in slurry transportation and technology. Publishes reports, proceedings and bibliographies.

UNITED MINE WORKERS OF AMERICA
900 15th Street
Washington, DC 20005
Telephone: (202) 638-0530

Description: Union of coal miners concerned with all aspects of coal mining and safety.

Index